BUOYANCY EFFECTS ON NATURAL VENTILATION

This book describes in depth the fundamental effects of buoyancy, a key force in driving air and transporting heat and pollutants around the interior of a building. This book is essential reading for anyone involved in the design and operation of modern sustainable, energy-efficient buildings, whether a student, researcher, or practitioner. The book presents new principles in natural ventilation design and addresses surprising, little-known natural ventilation phenomena that are seldom taught in architecture or engineering schools. Despite its scientific and applied mathematics subject, the book is written in simple language and contains no demanding mathematics, while still covering both qualitative and quantitative aspects of ventilation flow analysis. It is, therefore, suitable to both non-expert readers who just want to develop intuition of natural ventilation design and control (e.g., architects and students) and those possessing more expertise whose work involves quantifying flows (e.g., engineers and building scientists).

Dr. **Torwong Chenvidyakarn** is Senior Tutor and Lecturer in the Architectural Innovation and Management Program at Shinawatra International University, Thailand. He also currently works as a building consultant and publishes internationally. Previously, Dr. Chenvidyakarn was Chair of Technical Teaching in the Department of Architecture at the University of Cambridge, where he lectured on building physics, building innovation, architectural design, and environmental design in the Department of Architecture and Department of Engineering. He also held a full Fellowship in Architecture at Magdalene College and a Directorship of Studies in Architecture at Magdalene and Downing Colleges, Cambridge. He received the Happold Brilliant Award for Excellence in the Teaching of Building Physics in the Context of a Low Carbon Economy from the Chartered Institution of Building Services Engineers, UK. Dr. Chenvidyakarn carries out research in the area of sustainable buildings, with particular interests in the physics of natural ventilation, heating and cooling, and the impacts of climate change on the design and management of the built environment. A former researcher at the BP (British Petroleum) Institute in the United Kingdom, he has been a guest lecturer/design critic at a number of universities.

Buoyancy Effects on Natural Ventilation

Torwong Chenvidyakarn

*Former Fellow and Director of Studies in Architecture,
University of Cambridge, and Senior Tutor, Architectural
Innovation and Management Programme, Shinawatra
International University*

CAMBRIDGE
UNIVERSITY PRESS

CAMBRIDGE
UNIVERSITY PRESS

32 Avenue of the Americas, New York, NY 10013-2473, USA

Cambridge University Press is part of the University of Cambridge.

It furthers the University's mission by disseminating knowledge in the pursuit of
education, learning and research at the highest international levels of excellence.

www.cambridge.org
Information on this title: www.cambridge.org/9781107015302

First published 2013

Printed in the United States of America

A catalog record for this publication is available from the British Library.

Library of Congress Cataloging in Publication Data
Chenvidyakarn, Torwong.
Buoyancy effects on natural ventilation / by Torwong Chenvidyakarn.
    pages   cm
Includes bibliographical references and index.
ISBN 978-1-107-01530-2 (hardback)
1. Natural ventilation. 2. Buoyant ascent (Hydrodynamics)   I. Title.
TH7674.C47   2013
697.9'2–dc23   2013007957
ISBN 978-1-107-01530-2 Hardback

Contents

Preface   *page* ix
Acknowledgements   xv

1 Introduction   1

    1.1. The modelling quest   3
    1.2. Water-bath modelling   5
    1.3. The theoretical basis   9
    1.4. Applicability of water-bath modelling   14
    1.5. The cases examined   16

2 Some preliminaries 18

    2.1. Various conservation laws   18
        2.1.1. Conservation of mass   18
        2.1.2. Conservation of thermal energy   20
        2.1.3. Conservation of buoyancy flux   21
    2.2. Equilibrium and neutral level   23
    2.3. Bernoulli's theorem   26
    2.4. Effective opening area   27
    2.5. Application of the basic principles   30

3   Sources of identical sign   34

    3.1. Residual buoyancy   35
          3.1.1. Mixing ventilation   36
          3.1.2. Displacement ventilation   43
    3.2. The localised source   51
          3.2.1. Plume theory   51
          3.2.2. Sealed enclosure   58
          3.2.3. Ventilated enclosure   62
          3.2.4. Transient responses   69
          3.2.5. Multiple localised sources   78
    3.3. The distributed source   96
          3.3.1. Steady-state flow regime   98
          3.3.2. Evolution to steady state   103
    3.4. A combination of the localised source and the
          distributed source   116

4   Sources of opposite sign   126

    4.1. Flushing with pre-cooled air   127
    4.2. Pre-cooled ventilation of occupied spaces   143
          4.2.1. Cooling to above ambient air
                  temperature   144
          4.2.2. Cooling to below ambient air
                  temperature   156
    4.3. Maintained source of heat and internal
          cooling   165
          4.3.1. Distributed source of heat and distributed
                  source of cooling   169
          4.3.2. Localised source of heat and distributed
                  source of cooling   175
          4.3.3. Localised source of heat and localised
                  source of cooling   183

5 Some common flow complications arising from
   more complex geometries 190
   5.1. Openings at more than two levels   191
         5.1.1. Multiple stacks   191
         5.1.2. Multiple side openings   204
   5.2. Multiple connected spaces   218
         5.2.1. Multi-storey buildings   218
         5.2.2. Spaces connected sideways   230

Final remarks   247
References   249
Index   257

Preface

The reader may find it curious if I begin by saying that even after thousands of years of designing and constructing naturally ventilated buildings, our collective state of knowledge of one of the key forces driving their operation – buoyancy – still leaves much to be desired. I do not mean, of course, that a complete knowledge of buoyancy should have been obtained by now, for it would be unrealistic to expect that a knowledge of something this rich could be well and truly complete; I merely mean sufficient knowledge to enable us to design and operate buildings so that they deliver comfort and energy efficiency. Evidence abounds that today we, on the whole, still have not succeeded in acquiring such knowledge: there are many cases in which design/control schemes of natural ventilation systems are proposed or put in place that, even after considering the constraints associated with designing and constructing buildings, imply a lack of basic understanding of how buoyancy affects indoor air flows.

This problem, I believe, can be attributed in part to our preoccupation over the last century or so with mechanical ventilation and air conditioning, the two techniques of climate control that have become mainstream for modern buildings. Moreover, early work on natural ventilation focused not on buoyancy but on wind – it appeared to be the desire to obtain a knowledge of pressure distribution around a building subjected to incident wind and the influence of this on the rate of interior air flow that captured our attention at the time, not indoor air flow patterns or temperature structures arising from the effects of thermal buoyancy. Furthermore, and this is, in my view, the most critical factor of all, we have, by and large, been over-reliant on the knowledge of the working of buoyancy and rules of thumb for ventilation design and control that have been passed down from our forefathers, who derived them from their experience in traditional buildings. This traditional knowledge and intuition, although arguably serving generations of designers well in the past, have become increasingly inadequate in the face of modern buildings: the new kind of construction often possesses spatial characteristics, heating/cooling regimes, and occupancy patterns to which traditional knowledge and intuitions are not immediately applicable (such as an open-plan space with a lightweight fabric and a chilled ceiling). Our relative lack of scientific foundation in the area of buoyancy-driven flows has served only to limit our ability to tackle this new situation.

But it would be wrong to say that we have no material at all with which we may move forwards. A couple of decades or so back there was a surge in scientific research into the fluid mechanics of buoyancy-driven flows, as a consequence of a renewed interest in natural ventilation that first started in the 1970s in response to the oil crisis and became mature in the 1990s amidst the concern about global warming. This renewed interest in natural ventilation has led to insights being rapidly acquired into the working of buoyancy-driven flows, many of which have the ability to recalibrate – even revolutionise in some cases – our comprehension of the role that buoyancy plays on indoor air flow structures and temperature, thereby laying down a new basis on which effective natural ventilation strategies may be developed for modern buildings. Unfortunately, despite continued efforts to disseminate this knowledge, it has so far been predominantly confined within very limited circles of researchers, untaught in most architecture and engineering schools and unknown by most practitioners. As a result, natural ventilation remains largely an elusive climate control strategy, often mishandled by those involved in its design and operation.

It is my view that if natural ventilation is to be employed more effectively, and therefore make a real positive impact on the environment, the new insights need to be distributed more widely. This is the challenge that I try to tackle through this book: in writing it I aim to introduce to greater communities of architecture and engineering students, researchers, and practitioners the findings of various pieces of research which can provide an up-to-date grounding in natural ventilation design and control for modern buildings. I hope that, being equipped with this knowledge, my readers will find natural ventilation easier to handle, and see it as a viable alternative to mechanical ventilation and air conditioning.

However, as much as I would like it to be, this book is not exhaustive. The field of buoyancy-driven flows is so rich and expanding so rapidly in recent years that it is impossible to cover all that is relevant in a single volume. For this reason, I have selected topics that, I believe, are most essential. My experience in the building industry and university teaching has guided me in selecting these topics and how to treat them. I have worked in theoretical and physical modelling of air flows using small-scale laboratory experiments, and found the unique insights afforded by them indispensable for developing design and control strategies for actual buildings. It is no accident, therefore, that flow principles presented herein are largely those that have been acquired through this combination of investigation techniques, not just by me but also by my colleagues and other groups of researchers. Furthermore, I have found that many flow complications observed in real life can be clarified by first gaining an understanding of basic flow processes. Thus, the majority of the discussions in this book are based on spaces and sources of buoyancy of simple, generic geometry, which allow basic flow processes to be discerned more readily.

In structuring this book I have taken the liberty of classifying buoyancy-driven flows into two groups. One of these concerns a buoyancy source or sources of identical sign (as in a room containing either a source of heating *or* a source of cooling). The other concerns buoyancy sources of opposite sign (as in a room containing a combination of a source of heating *and* a source of cooling). Within each group, a number of problems are discussed that are treated largely independently from one another to allow the reader to home in on topics of his/her interest; cross referencing is, nevertheless, made where appropriate. A different structure could have been adopted, of course, but the aforementioned structure should be suitably simple while being capable of covering a good range of flow phenomena.

Two elements are combined in laying down arguments in this book. The first and arguably more important of the two is *qualitative discussions*. These are aimed at developing an intuition of the way air moves within a building, how it affects the interior conditions, and how it may be influenced by controlling parameters such as the location of the windows and the geometry of the space. This element is complemented by *quantitative formulae and rules*, which lay down more formally the relations between various flow properties (e.g., direction and volume flux) and the controlling parameters. Both of these elements are essential to making sound decisions on the design and control of natural ventilation systems and, therefore, should be considered in an integrated manner, where possible. Having said that, I have made the qualitative part quite self-sufficient, so that it may be approached on its own by non-specialist audiences or those less interested in flow quantification. The mathematics presented in this book is aimed at those possessing greater levels of expertise, and can give a more complete picture of flow, where required. However, it may be ignored without incurring serious penalty on the overall understanding. In addition, I have tried to use plain language throughout, with minimal jargon except where common technical terms are concerned with which readers will benefit from getting familiarised; in such cases brief descriptions are also provided for the terms.

Lastly, I would like to point out that it is the beauty of physics that it is universally applicable. This is true also for the physics of buoyancy-driven flows presented herein: the season or climatic zone within which a building operates is of no consequence to the general flow patterns or interior temperature structures described (although they can, and do, have an influence on whether certain ventilation techniques could or should be applied in certain circumstances). Through our discussions, it will be shown how an understanding of the physics of buoyancy lies at the heart of successful natural ventilation.

Torwong Chenvidyakarn

Acknowledgements

This book would not have been completed without the help of certain people. I would like, first of all, to thank Peter C. Gordon, Senior Editor at Cambridge University Press, Americas, for kind and helpful advice throughout the course of writing the book. In addition, I am grateful to Peter Grant, Daniel Godoy-Shimizu, and Nicola Mingotti for helping check the manuscript.

1 Introduction

Natural ventilation is the oldest strategy for ventilating buildings. Indeed, it had been the only one until the spread of fans and motors in the early twentieth century, which heralded the era of mechanical ventilation and air conditioning. At the time, thanks to cheap energy and increasingly polluted, noisy urban environments among other factors, mechanical ventilation and later full air conditioning were seen to be indisputably appropriate.

It was not until recently that the tide began to turn. Worldwide concerns about global warming and increasing energy costs have contributed to a decline in the popularity of full mechanical ventilation and air conditioning. This situation, coupled with the fact that in some cases these newer ventilation techniques have failed to provide clean air, resulting in 'sick building syndrome,' has also led to a growing revival of interest in natural ventilation. This changing trend in ventilation has been particularly noticeable in regions where ambient air can provide acceptable indoor temperatures, such as in temperate climates. The BedZED housing development

Figure 1.1. Houses at the Beddington Zero Energy Development (BedZED), Beddington, UK. Its rooftop cowls are used for natural ventilation.

in Beddington (Fig. 1.1), the Contact Theatre in Manchester (Fig. 1.2) and the Building Research Establishment office in Watford (Fig. 1.3) are just a few examples of modern buildings that use natural ventilation to great effect.

One aspect of natural ventilation that makes it particularly energy efficient is its ability to harness thermal buoyancy – a force that arises as a result of variation in the density of air subjected to heating/cooling and gravitational acceleration – for driving air and transporting heat and pollutants around the interior of a building. In fact, in modern buildings with tight construction and relatively high heat loads, buoyancy inputs from sources such as occupants, electrical equipment and ingress solar radiation alone can often provide sufficient

Figure 1.2. The Contact Theatre in Manchester, UK and its distinctive ventilation chimneys.

driving force to achieve thermal comfort and good indoor air quality. Moreover, thermal buoyancy can be combined with wind to enhance natural ventilation, enabling satisfactory internal conditions to be achieved more readily. Insights into the effects of buoyancy, then, are indispensable for successful natural ventilation design and control.

1.1. The modelling quest

However, it has not been easy to acquire insights into how buoyancy affects natural ventilation. Unlike the impact of wind, which may be explored relatively conveniently at a

Figure 1.3. The Building Research Establishment in Watford, UK. Here, natural ventilation is facilitated by solar chimneys located on the façade of the building.

small scale using building models placed in wind tunnels, the impact of buoyancy cannot be modelled physically at a small scale using air: the increased viscous effects at the smaller Reynolds numbers (the number that describes the relative effect of inertia and friction on a flow) obtained make the small-scale air flow dynamically dissimilar to the full-scale air flow, such that the former is not an accurate representation of the latter. As a result, in the past buoyancy-driven flows were studied mostly at full scale or close to full scale, using air. This was often inconvenient and costly, and so to circumvent the problem, in the 1990s a team at the University of

Cambridge developed a technique that used water (in combination with salt solutions) instead of air to model ventilation flows. Because water was now used as the medium fluid, Reynolds numbers were obtained at small scale that were highly comparable to those obtained at full scale. This allows the dynamics of the full-scale air flow to be captured by a small-scale water flow. Indeed, quantitative comparisons between flows in laboratories and full-scale measurements (by Lane-Serff 1989 and Savardekar 1990, for example) have confirmed that the small-scale water flow can accurately represent the large-scale air flow when convection is dominant and the flow is free of viscous and diffusive effects – the conditions that apply to many real-life ventilation situations. Because of the fluid medium it uses and the way the modelling system is set up (which is explained in more detail in Section 1.2), this modelling technique has become known as the *salt-bath* or *water-bath* technique.

1.2. Water-bath modelling

In setting up a water-bath experiment, a building model is constructed from transparent acrylic (or some other similar transparent material with relatively low thermal conductivity) at a scale ranging typically from 1:20 to 1:100. This model does not need to be an exact replica of the actual building it aims to represent; only the essential features controlling the ventilation, for example, the placement of the vents, must be incorporated accurately (Fig. 1.4). This model is then submerged in a large reservoir containing fresh water, which

Figure 1.4. Example of an acrylic building model used in water-bath modelling. The various openings of the full-scale building that the model represents are featured simply as holes drilled into the envelope of the model. In the figure, some of these holes are shown sealed with rubber stoppers; these may be removed during an experiment.

represents the ambient environment. Traditionally, salt solutions are injected into the model to represent heat inputs from localised sources such as radiators and concentrated groups of people, so that thermal buoyancy driving the ventilation is reproduced by density contrasts between the fresh water and the salt solutions. Because salt solutions are denser than fresh water, the re-created buoyancy acts downwards. Consequently, if salt solutions are used to represent warm air, the model must be inverted and viewed through an inverted camera so that they appear to rise within the model. Cold inputs can be replicated using water–alcohol mixtures that have densities less than that of fresh water. This arrangement is simply one of convenience because it avoids the need to add salt to the large body of the ambient reservoir. The resultant flow can be visualised using shadowgraph imagery (a technique whereby light is shone onto the flow from one side of the reservoir so that the flow's density field is captured as a shadow pattern on a screen placed on the other side of the reservoir). Dyes may also be added to the flow to highlight the flow pattern and the temperature structure, if needed, making the technique a highly intuitive means of capturing flow processes and of communicating between the design team and the client. Quantitative measurement of flow velocities may be made using digital image processing, and measurement of fluid densities may be performed using a refractometer or a conductivity probe.

Relatively recently, it has also been possible to model flows in a laboratory using 'real' (i.e., thermal) sources of heating/cooling, such as a heater or a chiller located within or attached to the building model. Buoyancy in this case is created by temperature differences between cool and warm water in the system, and calculations (e.g., by Gladstone & Woods 2001 and in the example given in Section 1.3) have confirmed that dynamic similarity to the full-scale flow can be obtained using this approach. This allows the effects of non-localised sources of buoyancy, such as a distributed body of occupants or a chilled ceiling, to be captured and studied. Furthermore, because warm water always rises and cool water falls, the model does not need to be inverted when using this setup. Flow visualisation may be done using shadowgraph imagery and dyes as in the salt system, and the temperature of the water may be monitored using thermocouples or thermistors connected to a computerised data processing unit.

Quantitative measurements have shown very good agreement between the flow structures achieved in hot water experiments and those achieved in salt-bath experiments, indicating that high Reynolds and Péclet numbers (the latter describes the ratio of the rate of advection of a physical quantity, such as heat and salt, to the rate of diffusion of the same physical quantity) are achieved in the water system as in the salt system (Linden 1999). This enables the two systems to be mixed within the same experiment, opening the door to modelling more complex flow phenomena such as those resulting from a simultaneous presence of a cluster of computer terminals and a chilled ceiling (e.g., in Section 4.3.2). Note that the effects of external wind forcing can also be modelled using the water-bath technique: flumes of controllable speed may be generated in a large reservoir of fresh water using a pump (see, e.g., Hunt & Linden 1997 and Lishman & Woods 2006).

At this point, it is useful to point out that other techniques are also available that may be used to investigate the effects of buoyancy, albeit not physically. Computational fluid dynamics (CFD) simulation and the numerical method are two examples of these. These techniques enable, to varying degrees, the examination and prediction of air flows within an enclosure, allowing ventilation quality and the thermal comfort level to be estimated. The details of these techniques and their applications are discussed in Awbi (1991), Etheridge and Sandberg (1996), Alamdari et al. (1998), Allard (1998) and Chen and Glicksman (2001), among others, and are not repeated here. Nor are the merits of these techniques compared with those of the water-bath technique; all techniques have their own strengths and weaknesses and are suitable for different circumstances, and readers are advised to exercise judgement as to their appropriateness on a case-by-case basis.

1.3. The theoretical basis

As mentioned earlier, the validity of the water-bath technique rests primarily on achieving dynamic similarity between the small-scale water flow and the full-scale air flow. The dimensionless parameters governing the dynamics of natural ventilation are the Reynolds number, Re, which describes the ratio of inertia to friction; the Péclet number, Pe, which describes the ratio of the rate of advection of a physical quantity, such as heat and salt, to the rate of diffusion of the same physical quantity; and the Rayleigh number, Ra, which describes the ratio of heat transfer by convection to that by conduction. The values of these three parameters achieved in a laboratory model must be comparable (though not necessarily identical) to those achieved in the full-scale building if dynamic similarity is to be satisfied.

In a buoyancy-driven flow system, the Reynolds, Péclet and Rayleigh numbers are defined in terms of the buoyancy force. This force can be described conveniently in terms of reduced gravity, g', which is the effective change in the acceleration of gravity acting on a fluid (water in the model and air in the full-scale building) of one density in contact with

a fluid of another density due to the buoyancy force, namely
(Morton, Taylor & Turner 1956):

$$g' = g\frac{\rho_0 - \rho}{\rho_0}, \qquad (1.3\text{-i})$$

where g is gravitational acceleration, and $(\rho_0 - \rho)$ is the difference in density between the fluid outside the system and the fluid inside (with ρ_0 being the density of the non-buoyant exterior fluid, usually taken to be the reference density). This relative density difference can, in turn, be related to the difference in temperature by the relation (Morton, Taylor & Turner 1956)

$$\frac{\rho_0 - \rho}{\rho_0} = \alpha \Delta T, \qquad (1.3\text{-ii})$$

where α is the volume expansion coefficient of the fluid (which describes how much the fluid expands as its temperature increases), and ΔT is the difference in temperature between the fluid inside the building and the fluid outside.

In treating buoyancy-driven flows in most building contexts (except in certain situations, such as where a strong fire is involved) it is customary to adopt the *Boussinesq approximation*. In doing so, we assume that the density variation $(\rho_0 - \rho)$ arising from heating/cooling is small compared with the reference density ρ_0, so that the effect of density variation on inertia is negligible. This allows the fluid density to be regarded as a constant in most expressions except when it appears in buoyancy terms; that is, when it involves gravity, because gravity is sufficiently strong to make the specific weights of hot and cold fluids appreciably different. When the Boussinesq approximation is adopted, variation in other fluid properties such as the volume expansion coefficient, kinematic viscosity (i.e., the fluid's stickiness) and diffusivity (i.e., the fluid's ability to transport a physical quantity such as heat and salt without bulk motion) is also usually neglected. These approximations obviously lead to a loss of accuracy in calculating the flow rate and temperature in the buoyant region. However, for the relatively small temperature differences usually involved in building natural ventilation flows, this loss of accuracy is generally acceptable: a change of about 10 K in the temperature of air, for example, will change the value of its volume expansion coefficient by only about 3%, though the approximations help simplify the calculation immensely.

Nevertheless, it must be noted that the Boussinesq approximation cannot be applied to all kinds of natural ventilation flow: certain flows – for example, ones associated with the ventilation of strong fires – involve considerable temperature variation, leading to non-negligible expansion or contraction in the volume of fluid, meaning that density may not be regarded as constant. Moreover, the behaviour of non-Boussinesq flows is fundamentally different from that of Boussinesq ones. For instance, the flux of buoyancy of a non-Boussinesq plume is not conserved, unlike that of a Boussinesq plume. Readers interested in treatment of non-Boussinesq flows are advised to consult Rooney and Linden (1996) and Lee and Chu (2003), for example.

For now let us concentrate on the Boussinesq flow. For such a flow driven by reduced gravity, it can be shown through dimensional analysis – an examination of the relations between physical quantities using only their dimensions, e.g., length, time, mass and volume – that the flow velocity scales on $(g'H)^{1/2}$. Applying this scaling, we may write the following for the Reynolds, Péclet and Rayleigh numbers, respectively:

$$Re = \frac{\left(\sqrt{g'H}\right) H}{\nu}, \qquad (1.3\text{-iii})$$

$$Pe = \frac{\left(\sqrt{g'H}\right) H}{\kappa}, \qquad (1.3\text{-iv})$$

and

$$Ra = \frac{g'H^3}{\kappa \nu}. \qquad (1.3\text{-v})$$

where H is the characteristic length scale of the flow – that is, the distance over which the flow develops, which is usually taken as a vertical one for a buoyancy-driven flow; ν is the kinematic viscosity of the fluid; and κ is the diffusivity of the fluid (this last constant refers to mass diffusivity in the case of salt-bath experiments and thermal diffusivity in the case of hot water experiments). Dynamic similarity is achieved because, although the length scale of the model is much smaller than that of the actual building, the kinematic viscosity and diffusivity of water and salt solutions are also smaller than those of air. Moreover, reduced gravity can be made much greater in the model than in the actual building when salt solutions are used, allowing dynamic similarity to be achieved more readily. Note that, in salt-bath experiments, in which there is no temperature difference, only the Reynolds and Péclet numbers need to be considered. However, in hot water experiments, in which buoyancy is created by temperature differences, it is the Reynolds and Rayleigh numbers that need to be considered (see Linden 1999 and Gladstone & Woods 2001). Typically, at both small and full scale, the values of the Reynolds and Péclet numbers achieved are in excess of 10^3, indicating that flows are essentially free of viscous and diffusive effects and that the transport of heat is accomplished primarily by turbulent convective fluxes (Holman 1997; Linden 1999). The values of the Rayleigh number in both scales usually exceed 10^4, signifying that heat transfer is accomplished chiefly by convection rather than by conduction (Holman 1997).

  As an example to illustrate the above principles, consider a building of height 10 m and with a temperature difference between the inside and outside of 10°C. The building is replicated using a 1:50 scale model of height 20 cm. A heat source in the model produces a temperature difference of, say, 15°C between the water in the model and the water in the reservoir. (This value of temperature difference is quite conservative; higher values are often achieved that allow dynamic similarity to be obtained more readily.) Let us denote the full scale by the subscript f and the model scale by the subscript m. The following typical values of constants may be taken in calculation: gravitational acceleration $g \sim 10$ m/s^2; the volume expansion coefficients $\alpha_f \sim 10^{-3}$ 1/K and $\alpha_m \sim 10^{-4}$ 1/K; the kinematic viscosities $\nu_f \sim 10^{-5}$ m^2/s and $\nu_m \sim 10^{-6}$ m^2/s; and the thermal diffusivities $\kappa_f \sim 10^{-5}$ m^2/s and $\kappa_m \sim 10^{-7}$ m^2/s. Using Eqs. (1.3-i) and (1.3-ii) we obtain $g'_f \sim 10^{-1}$ m/s^2 and $g'_m \sim 10^{-2}$ m/s^2. Eq. (1.3-iii) then gives $Re_f \sim 10^6$ and $Re_m \sim 10^4$. Further,

Eq. (1.3-v) gives $Ra_f \sim 10^{12}$ and $Ra_m \sim 10^9$. These comparably high magnitudes of Reynolds and Rayleigh numbers indicate that dynamic similarity is achieved between the small-scale water flow and the full-scale air flow.

To extrapolate the information obtained from the small-scale model to the full-scale building, the following general relations may be used (Linden 1999):

$$\frac{\text{Time}_f}{\text{Time}_m} = \frac{\sqrt{H_f g'_m}}{\sqrt{H_m g'_f}} \qquad (1.3\text{-vi})$$

$$\frac{\text{Velocity}_f}{\text{Velocity}_m} = \frac{\sqrt{g'_f H_f}}{\sqrt{g'_m H_m}} \qquad (1.3\text{-vii})$$

and

$$\frac{\text{Buoyancy flux}_f}{\text{Buoyancy flux}_m} = \frac{\sqrt{g'^3_f H^5_f}}{\sqrt{g'^3_m H^5_m}}. \qquad (1.3\text{-viii})$$

1.4. Applicability of water-bath modelling

At this point two questions may arise pertaining to water-bath modelling. First, how does the technique account for the effects of radiative and conductive heat transfer? Second, if these effects are not accounted for, how realistic are the results obtained from water-bath modelling?

These questions are two of those most frequently raised regarding the technique, and in addressing them it is important first to clarify that analysis using water-bath modelling is usually performed on flow situations in which convection is dominant, controlling the flow pattern and interior temperature structure, and conduction and radiation are of secondary influence. This is not to say that radiation and conduction can never have an impact on interior air flows – indeed there are cases in which radiant sources of heat or conductive losses through the fabric of the building modify the internal air flow pattern and the temperature structure substantially – it is just that a clearer and deeper understanding of the underlying mechanics of flows can be acquired by first ignoring these effects, and that water-bath modelling is a suitable investigative tool in this context. This may sound limiting, but in fact situations in which convection dominates are quite common in modern naturally ventilated buildings. In such buildings, conduction and radiation are often limited by the well-insulated fabric whereas air movement remains strong. These conditions allow the majority of heat gain from radiative and conductive sources (e.g., the occupants' bodies, light bulbs, heated blinds and the cold/warm surfaces of exposed thermal mass) to be converted into convective gain; the water-bath technique treats conductive and radiative heat transfer elements implicitly as part of an overall heat balance that leads to a net effective heat flux from a convective source.

The aforementioned treatment of the different modes of heat transfer is appropriate because the goal of water-bath modelling is not to replicate as realistically as possible the natural ventilation flow in a particular building – a point often misunderstood. Instead, the technique seeks to capture the dominant heat transfer process – that is, convection – with a view to observing the generic impacts of key ventilation

features, such as the window height or the size of the heat
load, on the flow structure and the interior temperature. From
these, insights into the fundamental flow mechanics and principles for control may be obtained. For this reason, unlike in,
say, CFD simulation and full-scale modelling, models used in
water-bath modelling are not exact replicas of actual buildings. Rather, they are analogues, containing only key ventilation features, as mentioned earlier. These analogue models,
though relatively simple, allow mathematical models to be
developed and tested that parameterise the flows. As shown
later in this book, once obtained, these mathematical models
can be scaled up or down and, with care, applied (by substituting appropriate values of variables into the models) to a
variety of practical flow situations, enabling flow-related processes and properties such as the transport of indoor pollutants, the rate of (convective) heat loss and the level of indoor
comfort to be quantified. Where necessary, the effects of comparatively small radiative and conductive heat transfer taking
place in specific buildings may be accounted for in the models
by coupling appropriate equations describing conductive and
radiative heat flows to the models derived from water-bath
modelling (see, e.g., Chenvidyakarn & Woods 2010).

1.5. The cases examined

It is results from investigations using a combination of water-
bath modelling and mathematical modelling that have fur-
nished material for this book. Therefore, flow situations
examined herein are ones in which convection is dominant. In

addition, we assume that the ambient environment in which each building is located is unstratified; that is, there is no discernible variation in the density of ambient air in the vertical direction, an approximation of the exterior environment typical in the building context. Also, we assume that none of the flows discussed herein is subjected to wind forces, in order to concentrate on the fundamental effects of buoyancy, which, I believe, should be understood before the more complex interaction between wind and buoyancy can be examined. This preclusion of the effects of wind naturally limits our discussion to indoor air flows. Finally, we regard air as an incompressible fluid, as is usually the case in the context of naturally ventilated buildings where no large pressure is concerned.

2   Some preliminaries

Before examining a particular flow it is useful to review some scientific preliminaries.

2.1. Various conservation laws

2.1.1. Conservation of mass

The conservation of mass is a powerful law that allows the amount of air flowing into and out of a building to be traced, thus enabling the ventilation rate to be estimated. The law of conservation of mass, when combined with the law of conservation of thermal energy and that of buoyancy flux, also allows the interior temperature structure to be determined. For a flow system whose physical boundaries – the walls, floor and ceiling – are all fixed during the ventilation process, the conservation of mass dictates that the amount of air entering the system at any given time equals the amount of air leaving it at the same time (assuming, as mentioned earlier, that air is incompressible):

$$M_{IN} = M_{OUT}, \qquad (2.1.1\text{-i})$$

where M has the dimension of mass per unit time. In this situation, if the rate of change in the volume of air due to heating/cooling is small compared with the overall ventilation volume flux, the Boussinesq approximation applies, allowing the density of air to be treated as a constant. Thus, the conservation of mass flux M may be approximated by the conservation of volume flux Q; that is,

$$Q_{IN} = Q_{OUT}; \quad M = \rho Q, \qquad (2.1.1\text{-ii})$$

where ρ is the density of air.

For a system whose physical boundaries change during the flow process, the mass flux of air through the system may be expressed by the differential equation

$$\frac{dM}{dt} = \rho \frac{dV}{dt}, \qquad (2.1.1\text{-iii})$$

where t is time and V is the volume of the system. Obviously, solving Eq. (2.1.1-iii) is more difficult that solving Eq. (2.1.1-ii). However, a simplification may be made that treats any smooth, continuous change in the boundary conditions as a series of small discrete changes. This is akin to saying that a window or a door is closed or opened suddenly, all at once or in little steps, rather than continuously. This helps simplify the analysis considerably: the simple mass balance equation (2.1.1-ii) may now be used, and the evolution of the interior temperature for each step change may be determined using a combination of Eqs. (2.1.2-i) and (2.1.2-ii) given in Section 2.1.2.

2.1.2. Conservation of thermal energy

Often invoked alongside the conservation of mass is the conservation of thermal energy. The first law of thermodynamics states that energy can be neither created nor destroyed; it can only change forms. Therefore, all energy gains and losses to and from a ventilated space have to be accounted for in determining the temperature inside the space. For a system of ventilation in which convective heat transfer is dominant, conductive and radiative heat transfer elements may be treated implicitly as part of a net effective heat flux from a convective source, as discussed in Section 1.4. In these conditions, the temperature achieved in the space at steady state may be conveniently determined from the balance between the total heat flux from the convective source, H_{TOTAL} say, and heat loss driven by ventilation H_V,

$$H_{\text{TOTAL}} = H_V = \rho C_P Q \Delta T \qquad (2.1.2\text{-i})$$

where Q is the ventilation volume flux, ΔT is the temperature difference between the air inside and outside the space, ρ is the density of air and C_P is the specific heat capacity of air (which describes how much thermal energy is required to change the temperature of air). As for a system in transience, the evolution of the temperature in the space may be expressed in differential form as

$$\rho C_P V \frac{d\Delta T}{dt} = H_{\text{TOTAL}} - H_V, \qquad (2.1.2\text{-ii})$$

where H_V is given by Eq. (2.1.2-i). These expressions may be coupled with expressions for conductive and radiative heat transfer in cases where the two components are significant, for example, when the space is poorly insulated, to build up a hierarchy of models for determining the temperature in the space.

2.1.3. Conservation of buoyancy flux

A conservation law specific to buoyancy-driven flows is that of conservation of buoyancy flux. This conservation law is often invoked in situations where the room is stratified and its interior temperature structure is to be determined (see, e.g., Section 3.2.5). Its derivation may be explained with reference to a Boussinesq flow rising from a source of heat flux H_W, similar to a thermal plume rising from a small radiator. Such a flow is driven by buoyancy flux B_0, which is a function of the reduced gravity g' produced by the source and the associated volume flux Q,

$$B_0 = g'Q = g\alpha\Delta T \frac{H_W}{\rho C_P \Delta T} = \frac{g\alpha H_W}{\rho C_P}, \qquad (2.1.3\text{-i})$$

where α and C_p are the volume expansion coefficient and specific heat capacity of the fluid, respectively. If the environment in which the flow develops is unstratified, $d\rho_0/dz = 0$, the ambient fluid carries no heat anomaly compared with the fluid in the flow. As a result, there is no removal or addition of heat from or to the buoyant convective parcel as it rises, and by virtue of the conservation of thermal energy, H_W is constant and the buoyancy flux is conserved with respect to height. In this way, the conservation of buoyancy flux may be regarded as an approximation of the conservation of thermal energy, as represented by a heat surplus in the buoyant region compared with the ambient environment. Morton, Taylor and Turner (1956) were among the first to make use of this principle in analysing the general behaviour of a thermal plume; we discuss their work in more detail in Section 3.2.1.

For a buoyancy-driven flow produced by a non-thermal source, such as a salt solution, the aforementioned principle also applies, except that, in this case, the concentration of the salt solution in the flow relative to that in the environment determines the buoyancy flux,

$$B_0 = g'Q = g\beta \Delta S Q, \qquad (2.1.3\text{-ii})$$

where β describes the relative change in density with salinity, and ΔS the salinity difference between the fluid in the flow and the ambient fluid (which can be measured as a mass fraction in the salt solution). In an unstratified environment, the ambient fluid contains no salt anomaly relative to the fluid in the flow, and so there is no removal or addition of buoyancy from or to the flow at all heights. Consequently, buoyancy flux is conserved.

For a flow in a stratified environment, the aforementioned principles apply, but buoyancy flux is conserved only within an environmental layer of the same density. It reduces as the flow rises into a more buoyant layer, due to less density contrast between the fluid in the flow and the ambient fluid. The behaviour of the flow as it crosses one layer into another may be quantified by adjusting the boundary conditions of the flow accordingly (see, e.g., Section 3.2.5 and Linden & Cooper 1996).

2.2. Equilibrium and neutral level

Often applied in conjunction with the conservation laws outlined in the preceding sections are the concepts of equilibrium and neutral level. Throughout this book we treat air as a homogeneous, inviscid (i.e., non-sticky), incompressible fluid. Such a fluid, when at rest, is in a state of neutral equilibrium; that is, the weight of each fluid element is exactly balanced by the pressure exerted on it by the neighbouring fluid elements. For natural ventilation flows driven purely by buoyancy, this pressure is hydrostatic and given by

$$P = -g \int_0^z \rho \, dz, \qquad (2.2\text{-i})$$

where g is gravitational acceleration, ρ is the density of air and z is the vertical distance (upward positive). The system is in equilibrium when the density and pressure are constant across the horizontal. This leads to stable stratification when heavier, colder air lies underneath lighter, warmer air; and to instability (technically called Rayleigh–Taylor instability) when heavier, colder air lies atop warmer, lighter air. In the latter circumstance, displacements of density occur across the horizontal to restore the stability, leading to convective motions. Such convective motions can be observed, for example, in a warm room into which relatively cold air is introduced at a high level (see, e.g., Section 3.1.1).

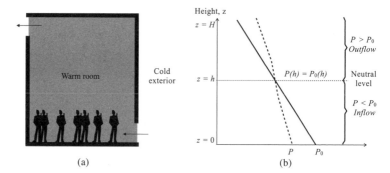

Figure 2.1. Neutral level and direction of flow in a warm room in a relatively cold environment.

Consideration of hydrostatic pressure is key to determining the direction of flow through an opening of a building. Take a warm room with vents at two levels located in a relatively cold, unstratified environment shown in Fig. 2.1a. The density of air outside the room, ρ_0, is constant, but the density of air inside, ρ, is a function of height, $\rho(z)$. Pressure gradients within and without the room along the vertical axis (denoted by P and P_0, respectively) can be described by Eq. (2.2-i), and these are shown schematically in Fig. 2.1b. Because the interior is warmer than the exterior, ρ is less than ρ_0, and the interior pressure gradient is less than that outside. If we let $P_0 = P_0(0)$ at $z = 0$, the base of the room, and $P_0 = P_0(H)$ at $z = H$, the top of the room, then there exists a level $z = h$ somewhere up the height of the room at which the pressure inside the room equals the pressure outside, $P(h) = P_0(h)$. Such a level is called the *neutral level*. Above the neutral level, the pressure inside the room, P, is greater than the pressure outside, P_0, and so there is outflow. Below the neutral level, the

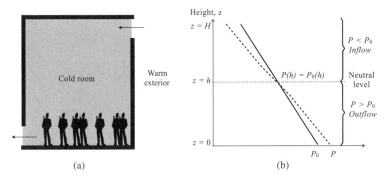

Figure 2.2. Neutral level and direction of flow in a cold room in a relatively warm environment.

pressure outside is greater than the pressure inside, and so there is inflow. Therefore, the overall ventilation through the room is upward.

In the case in which the room is kept cooler than the ambient environment (by means of thermal mass, for instance; Fig. 2.2a), the pressure profile will instead be as shown in Fig. 2.2b. In this case, the density of air inside the room, ρ, is larger than the density of air outside, ρ_0, making the interior pressure gradient greater than that outside. As before, the neutral level is located at a height at which the pressure inside equals the pressure outside, $P(h) = P_0(h)$. However, now above this level the pressure in the room is less than that outside, whereas below it the pressure in the room is greater. Thus, there is outflow through the lower opening and inflow through the upper opening, leading to a downward ventilation flow instead of an upward one.

Consideration of the neutral level in a manner similar to that described in Figs. 2.1 and 2.2 may be made for any kind of building subjected to any kind of heating/cooling. The principles are the same but the difficulty usually increases with the geometrical complexity of the space and the buoyancy source. This is particularly true for buildings with vents at more than two levels and those with stratified interiors, as will be seen in Chapter 5.

2.3. Bernoulli's theorem

The speed u of a flow through an opening of a building is related to the pressure P driving it by the relation

$$u = \sqrt{\frac{P}{\rho}}, \tag{2.3-i}$$

where ρ is the density of the fluid. To identify the pressure P along the streamline, Bernoulli's theorem may be used. Named after Swiss mathematician and physicist Daniel Bernoulli and derived from the principle of conservation of energy, this theorem essentially states that the sum of all forms of mechanical energy along the streamline of a steady flow of an incompressible fluid (such as air and water in our context) is constant at all points. In the context of buoyancy-driven flows this statement may be expressed by the relation

$$P + \frac{1}{2}\rho u^2 + \Delta\rho gh = \text{constant.} \tag{2.3-ii}$$

The first term of Eq. (2.3-ii) describes an initial pressure, the second term a change in kinetic pressure and the last term a change in hydrostatic pressure. For a buoyancy-driven flow through an enclosure, the definitions of the three terms are often as follows. The initial pressure is taken to be the atmospheric pressure exerted on the ambient fluid. The kinetic pressure is exerted as the flow comes into contact with a window, a grille or some other kind of opening on the enclosure, such as an entrance to a ventilation stack. The hydrostatic pressure arises from buoyancy associated with density differences between the fluid inside the enclosure and the fluid outside, $\Delta\rho$. The flow velocity u is related to the flow rate Q by the general relation

$$Q = uA^*, \qquad\qquad (2.3\text{-iii})$$

where A^* is the effective opening area whose definition is given in Section 2.4.

2.4. Effective opening area

In applying Eq. (2.3-iii) the natural contraction and expansion of the fluid (Fig. 2.3) as it enters and leaves an opening need to be taken into consideration. This contraction and expansion interrupts the smooth flow of the fluid, causing a loss in flow pressure. The sharper the edges of the opening are, the more abruptly the flow contracts and expands, and the more pressure is lost. Furthermore, a flow through a window or a doorway often encounters additional restrictions such as louvres, security bars and insect screens. These additional restrictions also hinder the flow, causing further pressure loss.

To take into account the effect of pressure loss on a ventilation flow, we usually express the net loss in terms of

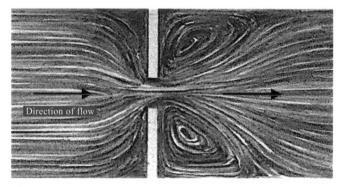

Figure 2.3. Sketch of flow streaks showing the sudden contraction and expansion of an inviscid fluid, such as air and water, as it passes through a sharp-edged constricted area, such as a window.

some coefficient c. The area of a single opening taking into account pressure loss across it can thus be expressed in terms of the effective opening area, A^*, as

$$A^* = ca, \qquad (2.4\text{-i})$$

where a is the visible opening area. The value of c approaches unity when the flow is smooth with little interruption or restriction (as in a flow through a Venturi pipe; Fig. 2.4), and approaches zero when the flow contracts or expands suddenly or is heavily restricted. For a steady flow of volume flux Q through a series of n openings, the total pressure loss across the entire system equals the sum of pressure losses across all the openings. Applying Bernoulli's theorem to this statement leads to

$$\frac{1}{A^*} = \frac{1}{\sqrt{2}} \left(\frac{1}{c_1^2 a_1^2} + \frac{1}{c_2^2 a_2^2} + \frac{1}{c_3^2 a_3^2} + \cdots + \frac{1}{c_n^2 a_n^2} \right)^{1/2}. \qquad (2.4\text{-ii})$$

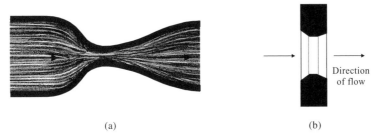

        (a)   (b)

Figure 2.4. The Venturi pipe (a) is shaped so that a fluid flowing through it contracts and expands smoothly, thereby losing very little pressure. The principle can be applied to a building by smoothing/chamfering the edges of a window to minimise sudden contraction or expansion of air flow (b).

It can be seen from Eq. (2.4-ii) that the effective opening area A^* is essentially restricted by the area of the smallest opening. For example, if a_1 approaches zero, then A^* also approaches zero. Recalling that the flow rate is proportional to the effective vent area (Eq. (2.3-iii)), we can see why, in real life, increasing the size of one opening once it is already larger than the other achieves little extra flow (Fig. 2.5), and that, in general, it is advisable to keep the sizes of all openings roughly similar to allow effective ventilation.

Note that for a flow through a long, narrow enclosure, for example, a slim ventilation stack, pressure loss due to friction in the flow path may be noticeable compared with pressure losses due to the expansion and contraction of the fluid across the openings of the enclosure. In this case, the coefficient c may be conveniently treated as an integrated constant accounting for both kinds of loss (e.g., Chenvidyakarn & Woods 2005a and Section 5.1.1).

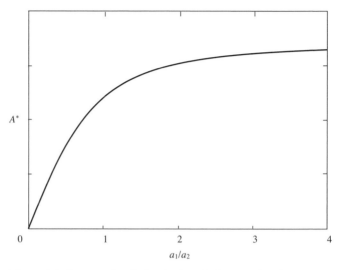

Figure 2.5. Impact of variation in the relative opening size on the effect-ive opening area. The plot assumes a system with two openings of areas a_1 and a_2 placed in series, with equal values of pressure loss coefficients, $c_1 = c_2$.

2.5. Application of the basic principles

The basic principles discussed in Sections 2.1–2.4 may be com-bined to gain a knowledge of flow-related quantities, such as volume flux and temperature, and, in turn, an entry into flow mechanics. As an example, let us consider the simple space in Fig 2.1a. Assume that the space is evenly heated by a source of flux H_H, and that its two vents are placed vertically apart by a distance H. The areas of the bottom and top vents are a_b and a_t, respectively, and pressure loss coefficients associated with these vents are c_b and c_t, respectively. At steady state, the room is heated to a temperature T while the outside air tem-perature is T_0. The room is in hydrostatic balance and the pres-sure profile in the room relative to that outside is essentially as shown in Fig. 2.1b. Using Bernoulli's theorem, pressure along the streamline may be traced from the inlet at the base of the room to the outlet at the top, giving

$$P_0 - \frac{1}{2}\rho_0 u_b^2 + (\rho_0 - \rho)gH - \frac{1}{2}\rho u_t^2 = P_0, \qquad (2.5\text{-i})$$

where P_0 is the atmospheric pressure acting on the ambient fluid, the term $1/2\rho_0 u_b{}^2$ is the kinetic pressure loss exerted as the flow passes through the bottom vent, the term $(\rho_0 - \rho)gH$ describes the hydrostatic pressure drive due to density deficiency in the space, and the term $1/2\rho u_t{}^2$ is the kinetic pressure loss exerted as the flow leaves the space through the top vent. The variable ρ is the density of the air inside the room, and ρ_0 is the density of the air outside. The variables u_b and u_t are the velocities of the flow as it passes through the inlet and outlet, respectively. The Boussinesq approximation allows $\rho = \rho_0$ in the kinetic term. Thus, on algebraically eliminating P_0 and rearranging the resultant equation, we have

$$\frac{1}{2}\rho_0 \left(u_b^2 + u_t^2\right) = (\rho_0 - \rho)gH, \qquad (2.5\text{-ii})$$

which effectively states that the total pressure loss in the system (left-hand side) equals the total pressure drive (right-hand side). The velocity of the flow through each vent, u_b and u_t, may be linked to its respective opening area by the relations (2.3-iii) and (2.4-i). Thus, Eq. (2.5-ii) becomes

$$\frac{1}{2}\left[\frac{Q_b^2}{c_b^2 a_b^2} + \frac{Q_t^2}{c_t^2 a_t^2}\right] = \frac{(\rho_0 - \rho)}{\rho_0}gH, \qquad (2.5\text{-iii})$$

where Q_b and Q_t are the inflow volume flux through the bottom vent and the outflow volume flux through the top vent, respectively. The term describing relative density deficiency on the right-hand side of Eq. (2.5-iii) can be related to the temperature excess in the space by the relation (1.3-ii) so that

$$\frac{1}{2}\left(\frac{Q_b^2}{c_b^2 a_b^2} + \frac{Q_t^2}{c_t^2 a_t^2}\right) = g\alpha\,(T - T_0)\,H. \qquad (2.5\text{-iv})$$

Mass conservation dictates that, at steady state, the mass flux of air into the space through the lower vent equals the mass flux of air leaving the space through the upper vent:

$$Q_b = Q_t = Q. \qquad (2.5\text{-v})$$

Therefore, it follows that

$$\frac{1}{2}Q^2\left[\frac{1}{c_b^2 a_b^2} + \frac{1}{c_t^2 a_t^2}\right] = g\alpha\,(T - T_0)\,H. \qquad (2.5\text{-vi})$$

The conservation of thermal energy relation (2.1.2-i) may be used to describe the temperature excess $(T - T_0)$ in terms of the heat flux H_H. This gives

$$Q^2\frac{1}{2}\left[\frac{1}{c_b^2 a_b^2} + \frac{1}{c_t^2 a_t^2}\right] = g\alpha\frac{H_H}{\rho C_P Q}H. \qquad (2.5\text{-vii})$$

Manipulating Eq. (2.5-vii) algebraically, we obtain an expression for the volume flux:

$$Q = A^{*2/3}\left[\frac{g\alpha}{\rho C_p}\right]^{1/3} H_H^{1/3} H^{1/3}, \qquad (2.5\text{-viii})$$

where A^*, the effective opening area, is given according to Eq. (2.4-ii) as

$$A^* = \frac{\sqrt{2}c_1 a_1 c_2 a_2}{\sqrt{c_1^2 a_1^2 + c_2^2 a_2^2}}. \qquad (2.5\text{-ix})$$

Furthermore, combining Eq. (2.1.2-i) with Eq. (2.5-viii), we have an expression for the interior temperature, namely

$$T = T_0 + \frac{H_H^{2/3}}{(\rho C_p)^{2/3} \, A^{*2/3} \, (g\alpha)^{1/3} \, H^{1/3}}. \qquad \text{(2.5-x)}$$

Examination may now be carried out on the flow mechanics, using, for example, Eqs. (2.5-viii) and (2.5-x) to explore how the flow rate and interior temperature change as certain ventilation parameters such as the effective size of the vents and the vent positions are varied. As will be seen in all of the following chapters, it is this kind of examination that leads to the acquisition of powerful principles of the general behaviours of a wide range of natural ventilation flows, which are central to the operation of naturally ventilated buildings.

3 Sources of identical sign

Spaces containing solely sources of heating *or* cooling can be said to contain sources of buoyancy of identical sign. In considering these spaces, we will take the motion of buoyancy as positive upward, as heat rises. The direction of motion is, however, irrelevant to the dynamics of flow, and sources of cooling may be treated as inverts of sources of heating, and vice versa.

Air movement driven by heating/cooling from sources of identical sign can take many forms, depending on how the sources behave within the space. The basic forms of flow, along with the associated temperature profiles, must be invoked in order to gain insights into frequently encountered phenomena, such as the feeling of cold feet in a heated room, the accumulation of smoke during a fire, and the presence of a region of high temperature at the top of a space such as an atrium subjected to heating. Transient flow processes must be given attention as well as steady-state flow regimes. The former has a direct implication for the control of a ventilation system when the timescale of evolution of the ventilation is comparable to or longer than the timescale of evolution of the heat load or occupancy. The latter is relevant when the ventilation evolves comparatively quickly.

We begin our discussion by looking at the flushing of a room containing no maintained source of heating/cooling but residual buoyancy, such as heat left after a period of occupancy. Two basic forms of natural ventilation flow are introduced, namely mixing and displacement, the appreciation of which will aid in understanding more complex flows discussed later on. Then we will look at the case in which the room contains a maintained source of heating/cooling. Two basic types of source are considered, as representing two ends of a geometrical spectrum. These are, at the one end, the localised source typified by a small radiator or a cluster of occupants in a relatively large space, and, at the other end, the distributed source typified by a chilled ceiling, a heated floor or a distributed body of occupants. Transitions between these extremes are also addressed. Finally, we end by examining the consequences of having a localised and a distributed source of heating/cooling in the same space.

3.1. Residual buoyancy

In an intermittently occupied space, such as a meeting room or a theatre, heat or contaminants left after a prolonged period of occupancy may lead to discomfort or poor indoor air quality. A common ventilation challenge is to remove these residual pollutants/heat to return the space to a state fit for occupancy. The time taken to do so is often intrinsically linked to the turn-around time or the pattern of occupancy of the space.

One technique that may be used for removing unwanted heat/contaminants from a space is to flush it with fresh air drawn from the exterior, assuming that the outside air is of acceptable quality. This, however, is not always straightforward and may lead to different flow regimes, depending primarily on the locations and number of vents and on whether or not the intake air has been pre-cooled/pre-heated. We deal with situations in which the intake air has been pre-cooled/pre-heated in Chapter 4, but for now let us look at situations in which air is taken unconditioned. Such situations may be modelled conveniently in a laboratory using a scaled enclosure filled with warm water submerged in a reservoir of cold water. The vents of the enclosure are sealed initially with rubber stoppers, which are later removed to start the flow. One of the following two basic flow regimes will be observed, namely *mixing* and *displacement* (Linden, Lane-Serff & Smeed 1990; Linden 1999).

3.1.1. Mixing ventilation

Mixing ventilation occurs when there is an opening or openings located at just one level in the space. This is what happens at the Hauser Forum seminar room at the University of Cambridge shown in Fig. 3.1, whose windows are located at the top of the wall. The essence of what takes place in this space and others similar to it may be described using the simple diagram in Fig. 3.2. Suppose that the room has a volume V and a uniform cross-sectional area S, and that its opening is of vertical dimension h_W and area A. At time $t = 0$, the temperature in

Figure 3.1. The Hauser Forum seminar room, University of Cambridge, with its high-level openable windows.

the space is uniform at T_0 and the exterior temperature is T_E, so that the ventilation is driven by initial reduced gravity $g'_0 = g\alpha(T_0 - T_E)$ (where α is the volume expansion coefficient of air). In these conditions, the neutral level lies between the

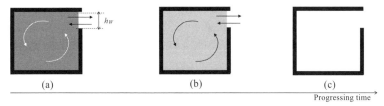

Figure 3.2. Transient mixing flow which develops when a warm room with an opening at the top is flushed with ambient air. (Following Linden, Lane-Serff & Smeed 1990.)

top and bottom edges of the opening (cf. Section 2.2). The opening therefore lets cool air into the space from the exterior through its lower part while allowing warm air to leave through its upper part. This creates a so-called *exchange flow* (so named because of the two-way flow) between the interior and exterior of the space. Once entering the space, the cool ambient air sinks and mixes with the original warm air (hence the name *mixing* ventilation). This mixing leads to weak stratification in the space and, for simple analysis, the space may be treated as well-mixed and having a uniform temperature (Fig. 3.2a). As the ventilation proceeds, the mixing causes the mean temperature T in the interior, and hence the reduced gravity g', to decrease towards those of the exterior air (Figs. 3.2a, b). The rates of evolution of the interior temperature and the reduced gravity are controlled by the volume flux Q of the exchange flow according to

$$\frac{dg'}{dt} = \frac{d\left[g\alpha\left(T - T_E\right)\right]}{dt} = -g'\frac{Q}{V}, \qquad (3.1.1\text{-i})$$

where the expression for Q may be obtained from tracing pressure along the streamline and applying Bernoulli's theorem. This gives

$$Q = A^*\left(g'h_W\right)^{1/2}, \qquad (3.1.1\text{-ii})$$

where the term $(g'h_W)^{1/2}$ describes the speed of the exchange flow. The effective opening area A^* is given according to Eq. (2.4-i) as $A^* = ca$, with c being the pressure loss coefficient and a the visible opening area. The evolution of the interior temperature and the reduced gravity during the flushing period

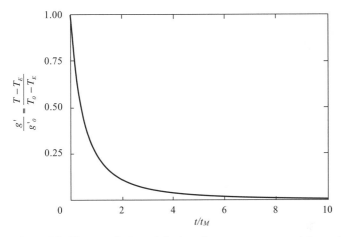

Figure 3.3. Time evolution of the interior temperature and the reduced gravity in a warm room with an opening located at a single level flushed with ambient air. (Following Linden, Lane-Serff & Smeed 1990.)

may be obtained from combining Eq. (3.1.1-i) with Eq. (3.1.1-ii). The resultant differential equation may be solved to give

$$\frac{g'}{g_0'} = \frac{T - T_E}{T_0 - T_E} = \left(1 + \frac{t}{t_M}\right)^{-2}; \quad t_M = \frac{2V}{A^* \left(g_0' h_W\right)^{1/2}},$$

$$(3.1.1\text{-iii})$$

where t_M is the mixing timescale. The result of this equation is plotted in Fig. 3.3.

Eventually, when all the residual air is drained from the space and the interior temperature becomes identical to that of the exterior, $T = T_E$, the reduced gravity decreases to zero, $g' = 0$, and the flow ceases, $Q = 0$ (Fig. 3.2c). Equation (3.1.1-iii) shows that the time taken for the room to drain depends on its volume, the effective opening area, the initial

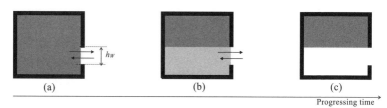

Figure 3.4. Transient mixing flow that develops when a warm room with an opening located at a lower level is flushed with ambient air. (Based on Linden 1999.)

reduced gravity and the vertical dimension of the opening. A larger room with a smaller opening and a lower initial interior temperature will have a longer mixing timescale t_M and therefore take longer to drain (Fig. 3.3). The turn-around time of occupancy of this room will therefore also be longer than that of a smaller room with a larger opening and a higher initial temperature.

As for a space whose opening is not located at the top but at some height lower down in the wall (Fig. 3.4), the transient exchange flow described earlier in this section still takes place, but only until the layer of cool ambient air rises to the top edge of the opening, after which the ventilation ceases (Figs. 3.4b, c; Linden 1999). The flow in this case can still be described with Eqs. (3.1.1-i)–(3.1.1-iii), but in applying these equations the volume V must now be taken as the lower, ventilated portion of the space only – that is, from the floor level to the level of the top edge of the opening. This reduction in the ventilated volume leads to a shorter draining time. This, on the face of it, may seem beneficial in reducing the turn-around time of the space, leading to greater financial profit in the case where the space is used for commercial purposes, for example, as a cinema. However, it must be noted that, with an opening located at a lower level, when the ventilation ceases, some residual heat/contaminants are left in the portion of the space above the opening. This pocket of warm air/contaminants, if lying within the occupied zone, could lead to discomfort and/or poor indoor air quality. For this reason, locating the opening lower down in the wall is generally inadvisable, except when it is positioned above the occupied zone.

It is useful to note that the draining of the space still takes place when the opening is non-vertical, for example, when the side window in Fig. 3.2 or 3.4 is replaced by an openable skylight. However, in this case, the flow that develops across the opening is quite different from that described in Figs. 3.2 and 3.4. As mentioned earlier, with a vertical opening, the flow is density driven, and the preferred arrangement between the inflowing cold ambient air and the outflowing warm residual air is one in which the warm air flows above the cold air, allowing a stable horizontal interface to establish between the two fluids. Owing to this separation of fluid, there is relatively little mixing between the cold and warm air at the opening. In this situation, if the opening is symmetrical about a horizontal axis (as in an opening located in the middle of a wall or a doorway stretching from floor to ceiling, for instance), the neutral level is established at the mid-height of the opening; and the values of the coefficient of pressure loss c range from 0.2 to 0.25 for sharp orifices such as most doors and windows (Brown & Solvason 1962a, b; Shaw & Whyte 1974; Linden & Simpson 1985; Lane-Serff, Linden &

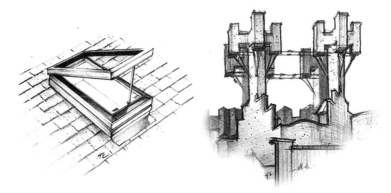

Figure 3.5. An opening on a sloping roof (left) and the horizontal openings of the ventilation chimneys at the Contact Theatre, Manchester, UK (right).

Simpson 1987). However, if the opening is asymmetrical, for example, as in a doorway flush with the floor but with a soffit to the ceiling, the interface is not at the mid-height of the opening, and the flow rate is modified accordingly (Dalziel & Lane-Serff 1991).

However, if the openings are non-vertical, as in those on a sloping roof and ventilation chimneys (Fig. 3.5) for instance, the form of exchange flow that develops is more complex. Now there is no preferred arrangement between the inflowing ambient air and the outflowing warm air, and the interface between the two fluids at the level of the opening is subject to Rayleigh–Taylor instability (i.e., an instability resulting from a heavier fluid being pushed from below by a lighter fluid). Consequently, there is significant mixing at the opening between the ambient air coming into the space and the warm air leaving it as the former flows downwards and the latter flows upwards. Dimensional analysis suggests that Eq. (3.1.1-ii) still applies in this case, but the value of the pressure loss coefficient c is now dependent on the angle θ that the opening makes with the horizontal, $c = c(\theta)$. For $\theta = 0°$, corresponding to the case of a horizontal opening, $c = 0.055$ is obtained if the opening is circular (Epstein 1988), and $c = 0.051$ is obtained if the opening is square, with its diagonal taken to be its width (Brown, Wilson & Solvason 1963). The mixing between the inflowing and outflowing fluids remains strong and the value of c remains constant up to the value of $\theta \approx 4 - 5°$, beyond which the value of c increases with that of θ, reaching a constant value of $c = 0.2$ when $\theta \geq 20°$ (Keil 1991; Davies 1993). At this point, the two-layered flow structure usually observed across a vertical opening is established, and the mixing between the inflowing and outflowing fluids becomes negligible. Comparing the values of c for non-vertical openings with those for vertical openings, we can see that vertical openings are more effective in ventilating a space, allowing about four to five times more volume flow than horizontal openings. This makes a space with a vertical opening drain significantly faster and have a shorter turn-around time than a comparable space with a horizontal opening.

3.1.2. Displacement ventilation

A different flushing flow structure altogether develops if the space in Fig. 3.2 has openings at two levels instead of one (Linden, Lane-Serff & Smeed 1990). Let us assume, for the sake of discussion, that these openings are located near the top

Figure 3.6. The atrium of the Evelina Children's Hospital, London, UK, has openings near the base and the crown, which allow flushing of the space using displacement ventilation.

and the base of the space, as in the case of the atrium of the Evelina Children's Hospital in London shown in Fig. 3.6. The simplified diagram of the flow that develops in this building is shown in Fig. 3.7. It may be seen that the initial temperature difference between the inside and outside of the building

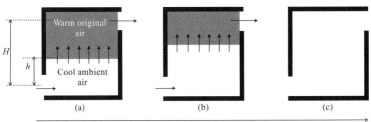

Figure 3.7. Transient displacement flow that develops when a warm room with openings at two levels is flushed with ambient air. (After Linden, Lane-Serff & Smeed 1990.)

draws cool ambient air into the interior through the lower opening. This air then displaces warm residual air in the building upwards and out through the upper opening (Fig. 3.7a). Owing to this movement of air, this form of flushing is called *displacement* ventilation. In this flow regime, if the vertical dimension h_W of each opening is sufficiently small compared with the vertical separation H between the openings, pressure gradients along the height of each opening are negligible. Consequently, the flow through each opening is uni-directional, and there is no significant mixing between the incoming ambient air and the warm residual air at the point of inflow. The ambient air, therefore, forms a layer at the base of the space upon entering it. This layer is separated from a layer of warm original air above by an interface (Fig. 3.7a). This interface is stable because the lighter, warm residual air lies atop the denser, cool ambient air. At the interface, there is a small transition region across which temperature varies smoothly between the adjacent layers, leading to the blurring of the interface. This transition region, which occurs in all natural

ventilation flows involving internal stratification, is only to be
expected because at the interface there is a great change in
temperature. However, the degree of blurring of the interface
varies from case to case, depending greatly on whether there is
any disturbance at the interface (e.g. one caused by penetrative convection discussed in Section 3.3.2). For the present case
where there is no disturbance at the interface, vertical mixing
between the two fluid layers is minimal, and for simplicity, the
interface may be treated as sharp. This simplification allows
the flow to be regarded as driven by reduced gravity acting
over the clearly defined warm upper layer, without altering
the overall flow picture. In these conditions, if we let h be the
height of the interface as measured from the mid-point of the
lower opening, we immediately obtain from Bernoulli's theorem the volume flow rate as a function of the initial reduced
gravity g'_0 and the height of the buoyancy column $(H - h)$,
namely

$$Q = A^* \left[g'_0 \left(H - h \right) \right]^{1/2}, \qquad (3.1.2\text{-i})$$

where A^* is the effective opening area given according to
Eq. (2.4-ii). The value of the coefficient of pressure loss c
across the openings is around 0.7 for sharp orifices such as typical windows and doors. Note that this value of pressure loss
coefficient, which is achieved when there are uni-directional
flows through the openings, is different from those achieved
when there are exchange flows; in the latter case the values of
c are typically much lower, between 0.05 and 0.25, depending
on the orientation of the openings (see Section 3.1.1).

As the flow progresses, more warm air is vented through the upper opening while more ambient air is drawn into the space through the lower opening. This leads to a depletion of the upper layer and a corresponding deepening of the lower layer (Figs. 3.7a, b). The interface between these two layers thus ascends at a rate that balances the net flow rate Q per unit cross-sectional area S of the room,

$$\frac{dh}{dt} = \frac{Q}{S}. \tag{3.1.2-ii}$$

Combining Eq. (3.1.2-i) with Eq. (3.1.2-ii) and solving the resultant differential equation, we obtain the evolution of the interface height as a function of time t:

$$\frac{h}{H} = 1 - \left(1 - \frac{t}{t_E}\right)^2; \quad t_E = \frac{2SH}{A^* \left(g'_0 H\right)^{1/2}}, \tag{3.1.2-iii}$$

where t_E is the draining timescale. Eventually, when the interface rises to the level of the upper opening, the ventilation ceases. At this point, the portion of the interior space from the lower opening to the upper opening attains the same temperature as that of the exterior air (Fig. 3.7c). Equation (3.1.2-iii) shows that the time taken to drain the space depends on the initial reduced gravity, g'_0, the cross-sectional area S of the space, the vertical separation H between the openings, and the effective opening area A^*. A larger and taller room (thereby having a larger t_E) will take longer to drain whereas a room with a larger effective opening area and a greater initial interior temperature (and hence a smaller t_E) will drain more swiftly (Fig. 3.8). Note that, as in the case of mixing ventilation discussed in Section 3.1.1, if the upper opening is not located

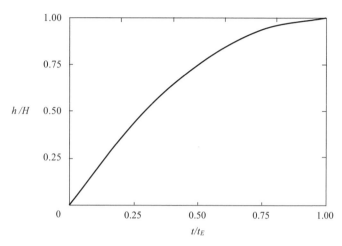

Figure 3.8. Time evolution of the interface height in a warm room with openings at two levels flushed with ambient air. (Based on model of Linden, Lane-Serff & Smeed 1990.)

at the very top of the space – for example, there is a soffit to the ceiling – there will be residual heat left in the portion of the space above the upper opening after the ventilation has ceased. This left-over heat could lead to discomfort if it sat within the occupied zone.

From the above discussion it is not difficult to appreciate that, for the same initial temperature difference and comparable room geometries, displacement ventilation allows more rapid ventilation and faster removal of residual heat/contaminants than mixing ventilation. This is mainly because, in displacement ventilation, the reduced gravity g' does not decrease over time but remains at its initial value throughout the ventilation process, due to the minimal mixing between the incoming ambient air and the outgoing residual air. In addition, the buoyancy head $(H - h)$ driving displacement ventilation is always greater than the head h_W driving mixing ventilation (except when the interface rises to the level of the upper opening).

This comparative performance of the two modes of flushing has important implications for the selection of the ventilation system. To use a single example, consider the flushing of a gas leak in a building. In this situation, the usual goal is to remove the leaked gas from the building as quickly as possible. Suppose that the gas is lighter than air. If mixing ventilation is used, there will be largely uniform mixing between the gas and the air in the room. The gas–air mixture will occupy the space from the base to the top edge of the vent, covering what is likely to be the occupied zone. As the ventilation progresses, the concentration of the gas will progressively reduce. However, this may take a long time before the level of concentration of gas reaches a safe level, depending on the size of the vent. On the contrary, if displacement ventilation is used, there will be relatively little mixing between the gas and the air. Instead, the gas will be quickly displaced upwards by the incoming fresh air and vented out through the top of the building. The ratio of the timescales of the two modes of ventilation as given by Eqs. (3.1.1-iii) and (3.1.2-iii) is $t_M/t_E \approx c_{(\text{mixing})}^{-1}(H/h_W)^{1/2}$. Because usually the vertical dimension of the vent h_W is a great deal smaller than the height H of the room and $c_{(\text{mixing})}$ ranges from about 0.05 (for horizontal vents) to 0.25 (for vertical vents), we have $t_M \gg t_E$, indicating that displacement ventilation is much more effective in flushing the hazard. A typical room 3 m in height with 30 cm tall vents, for example, would drain more than 10 times faster in a displacement regime than in a mixing regime.

In the case in which the leaked gas is heavier than air, the inverts of the aforementioned draining processes take place. With openings at two levels, ambient air enters the higher opening and the gas is drained through the lower opening. With one opening, if the opening is made at the base of the space, fresh air enters through the upper part of the opening while the gas is drained through the lower part. If the opening is made at a high level, the gas will not be drained but accumulate in the lower part of the room. This latter positioning of the vent, then, is not conducive to good ventilation.

A point may be noted: the discussion so far may have led to the perception that displacement ventilation is invariably more desirable than mixing ventilation. However, a pause for thought will lead to a rebuttal of this view: the choice of ventilation regime depends on the objective of ventilation. If the ventilation is done to remove contaminants/excess heat as quickly as possible from the space, as in the above examples, then displacement ventilation is clearly a better choice. However, if the ventilation is done in order to obtain minimum fresh air while containing and distributing heat within the space to achieve thermal comfort, as may be the case in winter, then mixing ventilation is more appropriate. In this latter case, trickle vents (i.e., very small openings) installed at high levels may be used to facilitate mixing flows while providing background ventilation.

3.2. The localised source

Having considered the transient flushing of a space containing residual buoyancy, we now turn our attention to situations in which a steady source of buoyancy is present in the space. These situations are of particular relevance when people live or work in a space for an extensive period. The first class of problems to be considered pertains to the localised source of buoyancy. This kind of source may be encountered in the form of a person, a concentrated group of occupants in a relatively large space, a radiator or a cluster of printers in an office. To analyse the buoyancy flux-carrying convective element of this kind of source, the plume theory developed by Morton, Taylor and Turner (1956) can be used. An examination is herein made of how the source interacts with the enclosure of the space and other kinds of source. We will see that dimensional analysis is a powerful tool for gaining an understanding of plume properties, and that the results of this approach can be combined with numerical parameters obtained experimentally to allow quantitative prediction of the ventilation flow and the temperature structure.

3.2.1. Plume theory

When a localised source of heating or cooling is present in a space, it creates a thermal plume. If the source is one of heating, the plume rises, whereas if the source is one of cooling, the plume sinks. Such a plume stays laminar (i.e., tidy in its structure) for only a short distance above the source (or

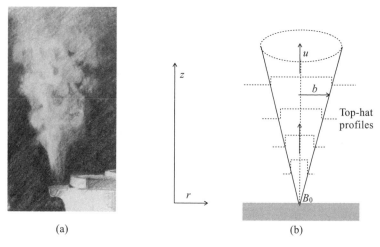

(a)   (b)

Figure 3.9. An example of a thermal plume (as generated by a boiling kettle) (a) and its simplified top-hat diagram (b).

below, in the case of a source of cooling) and soon becomes unstable and turbulent (i.e., chaotic in its structure). It is the behaviour of this turbulent plume that is fundamental to the overall ventilation of the space.

Schmidt (1941) was probably the first person to observe that a turbulent plume of hot air rising from a localised source tends to be confined within a conical region, as sketched in Fig. 3.9. The behaviour of such a plume is controlled chiefly by the process of entrainment (i.e., engulfing) of ambient air into the plume. Townsend (1970) gave a detailed theory of the mechanism of entrainment, but it suffices for our present purposes just to note that the turbulent plume has a sharp boundary that separates nearly uniform buoyant air within the plume from the surrounding air. This boundary is indented by large eddies that entrain the surrounding air into the plume as it rises. Provided that the motion of the plume is of sufficiently high Reynolds number, that is, $Re = ub/v > 10^3 - 10^4$ (where u is the vertical velocity of the plume, b the radial length scale and v the kinematic viscosity of the plume fluid), the overall behaviour of the plume is free of the influence of its kinematic viscosity and diffusivity. Also, the Reynolds number does not enter directly into the determination of the plume's overall properties. Thus the parameters governing the behaviour of the plume may be taken as comprising the pure (i.e., no mass) buoyancy flux B_0 at the source, the reduced gravity g', the height z and the radius r. These parameters may be defined with reference to cylindrical polar coordinates (z, r) as (Turner 1973)

$$B_0 = 2\pi \int_0^\infty ug'r \, dr, \qquad (3.2.1\text{-i})$$

where $g' = g\Delta\rho/\rho_0 = g(\rho_0 - \rho)/\rho_0$ is the plume-driving reduced gravity associated with density deficiency in the plume relative to the environment of density ρ_0. The velocity and density of the plume fluctuate, but the time-averaged profiles of velocity and density across the plume width may be taken as similar at all heights. Experimental measurements, for example, by Rouse, Yih and Humphreys (1952), showed that these mean plume profiles follow roughly Gaussian curves: the velocity and density deficiency in the plume decay radially (although the widths of the velocity and density fields are usually different). However, a simpler top-hat shape (Fig. 3.9b), which assumes that the plume has a uniform velocity and density across its width at a given height and zero velocity and density deficiency outside its boundary, may be adopted without affecting the overall behaviour of the plume (see, e.g., Worster & Huppert 1983; Linden, Lane-Serff & Smeed 1990 and Bower, Caulfield, Fitzgerald & Woods 2008). Applying this simplification and performing dimensional analysis, we immediately obtain that the reduced gravity and velocity of the plume decay above the source according to

$$g' \sim B_0^{2/3} z^{-5/3} \qquad\qquad (3.2.1\text{-ii})$$

and

$$u \sim B_0^{1/3} z^{-1/3}, \qquad\qquad (3.2.1\text{-iii})$$

while the radius of the plume increases with height as

$$r \sim z. \qquad\qquad (3.2.1\text{-iv})$$

Furthermore, combining Eq. (3.2.1-iii) and Eq. (3.2.1-iv), we obtain that the plume volume flux increases rapidly with height but slowly with the source's buoyancy flux according to

$$ur^2 \sim B_0^{1/3} z^{5/3}. \qquad\qquad (3.2.1\text{-v})$$

Equation (3.2.1-v) implies that the velocity of entrainment is related to the plume height (so that the volume of the plume increases as it rises). Thus the process of entrainment may be modelled as a radially inward horizontal velocity at the effective periphery of the plume ($r = b$ say), with this horizontal velocity taken as some fraction ε of the plume vertical velocity u at any level (Batchelor 1954). Applying this entrainment assumption, the top-hat plume profiles mentioned earlier and the Boussinesq approximation (which is valid when density variations are so small that they do not affect the plume inertia), we obtain the following equations from the conservation of mass, momentum and buoyancy, respectively:

$$\frac{d}{dz}(\pi b^2 u) = 2\pi \varepsilon b u, \qquad (3.2.1\text{-vi})$$

$$\frac{d}{dz}(\pi b^2 u^2 \rho_0) = \pi b^2 g \Delta\rho \qquad (3.2.1\text{-vii})$$

and

$$\frac{d}{dz}(\pi b^2 u g \frac{\Delta\rho}{\rho_0}) = \pi b^2 u \frac{g}{\rho_0} \frac{d\rho_0}{dz}, \qquad (3.2.1\text{-viii})$$

where, for boundary conditions, we assume that the radius, mass and momentum flux of the plume at the level of the source ($z = 0$) are zero, and that its density deficiency is released at a constant rate. For an unstratified environment, we have $d\rho_0/dz = 0$, and the right-hand side of Eq. (3.2.1-viii) vanishes, giving the result $b^2 u g \Delta\rho/\rho_0 = B_0/\pi = B$ say, showing that the flux of buoyancy in the plume is constant at all heights (i.e., the buoyancy flux is conserved; see Section 2.1.3). The density difference may then be eliminated from Eq. (3.2.1-vii) provided that the plume remains narrow, and similarity solutions obtained for the values at height z of the plume's momentum flux, $M_{(z)}$, velocity $u_{(z)}$, reduced gravity $g'_{(z)}$ and radius $b_{(z)}$ namely:

$$M_{(z)} = \left(\frac{9}{10}\varepsilon B_0\right)^{2/3} z^{4/3}, \qquad (3.2.1\text{-ix})$$

$$u_{(z)} = \frac{5}{6\varepsilon}\left(\frac{9}{10}\varepsilon B\right)^{1/3} z^{-1/3}, \qquad (3.2.1\text{-x})$$

$$g'_{(z)} = \frac{5}{6\varepsilon}\left(\frac{9}{10}\varepsilon B\right)^{-1/3} Bz^{-5/3}, \qquad (3.2.1\text{-xi})$$

$$b_{(z)} = \frac{6}{5}\varepsilon z. \qquad (3.2.1\text{-xii})$$

Combining Eq. (3.2.1-xi) with Eq. (2.1.3-i) and substituting B with B_0/π leads to an expression for the volume flux of the plume at height z, namely

$$Q_{P(z)} = \lambda B_0^{1/3} z^{5/3}, \qquad (3.2.1\text{-xiii})$$

where $\lambda = (6\varepsilon/5)(9\varepsilon/10)^{1/3}\pi^{2/3}$. For a localised source of heat flux H_P the variable B_0 is defined according to Eq. (2.1.3-i) as

$$B_0 = \frac{g\alpha H_P}{\rho C_P}, \qquad (3.2.1\text{-xiv})$$

with g being gravitational acceleration, α the volume expansion coefficient of the plume fluid, ρ the density of the plume fluid, and C_P the specific heat capacity of the plume fluid. Eqs. (3.2.1-ix)–(3.2.1-xiii) show that at the level of the source, $z = 0$, the momentum flux, volume flux and radius of the plume are zero but its velocity and reduced gravity are large. Actual sources, however, usually have non-zero dimensions (even a very small radiator or a person), and so the volume flux of the plume from such a source is usually estimated by assuming that there is a *virtual* source of zero dimension located at a distance z_0 below the level of the visible source. This distance z_0 is determinable from data collected from full-scale calibration experiments using the actual source in question. Taking into account this virtual source, Eq. (3.2.1-xiii) may be rewritten as

$$Q_{P(z)} = \lambda B_0^{1/3} (z + z_0)^{5/3}, \qquad (3.2.1\text{-xv})$$

where the distance z_0 can be determined from the relation

$$Q_{P(0)} = \lambda B_0^{1/3} z_0^{5/3}. \qquad (3.2.1\text{-xvi})$$

In the case of a plume rising through a stratified environment, for example, hot air rising from a radiator in a room in which residual heat accumulates in the upper part of the space, the buoyancy flux of the plume remains constant so long as the plume rises through a layer of environmental fluid of the same density, but reduces as the plume crosses from a layer of denser fluid to a layer of lighter fluid. Approaches for modelling the behaviour of the plume in this situation have been discussed by Morton, Taylor and Turner (1956); Turner (1973); Cooper and Linden (1996); and Bower, Caulfield, Fitzgerald and Woods (2008), among others. Essentially, the behaviour of the plume in the lighter layer may be considered that of a plume rising from an areal source of finite volume flux and momentum flux located at the interface between the two layers, and developing in an environment of uniform density. This areal source, in turn, may be modelled as originating from a virtual source of some smaller finite momentum flux but zero volume flux located some distance below the interface.

We emphasise here that the plume considered thus far
is one which may be classed as purely thermal; that is, the
plume carries negligible, or no, mass flux. In cases where a
finite, appreciable mass flux is involved, for example, strong
cold air from an air-conditioning unit, a heavy gas leak or
hot air released from certain kinds of floor diffusers, the
mass flux of the plume may lead to an overall flow pattern
and interior temperature structure that are quite different
from those achieved with the pure thermal plume, depending
primarily on the mass flux of the plume compared with the
building-scale ventilation flux. This problem is discussed in
the context of an air-conditioned space by Woods, Caulfield
and Phillips (2003), for example.

3.2.2. Sealed enclosure

The plume theory outlined in Section 3.2.1 allows us to examine flows driven by a localised source in an enclosed space,
such as a radiator or a person in a room. The first of these
flows that we will look at is one developed in a space with
no opening, equivalent to a room with all its doors and windows firmly shut. The behaviour of the plume in this situation
can be modelled experimentally using either a salt solution or
warm water. If a salt solution is used, the salinity of the plume
relative to that of the reservoir fluid controls the buoyancy
flux. The salt content in the plume fluid, measured as a mass
fraction in solution, typically ranges from about 3% to 10%.
Within this range, the Boussinesq approximation applies, and
a constant value of $\beta \approx 0.007$ may be taken to describe the

relative change in density with salinity (Eq. (2.1.3-ii); Liver-more & Woods 2008). If warm water is used, care should be taken to control the temperature of the plume to within relatively small values above that of the ambient environ-ment in order to allow the Boussinesq approximation to be applied (and thus avoid complications in calculation). Regard-less of the technique used, the plume must be injected at low momentum so as to allow the ventilation to be driven primar-ily by density differences rather than being forced. Also, the volume flux of the plume at the source must be kept very small compared with the volume flux driven by the overall ventil-ation (typically below 5%) in order to imitate the zero-mass thermal plume in real life.

Baines and Turner (1969) and Worster and Huppert (1983) showed how such a purely thermal, Boussinesq plume behaves in a sealed space. If the plume originates from the base of the space, it rises to the top, entraining on its way the surrounding cold air within the space. On hitting the ceil-ing, the plume spreads out horizontally, and then descends along the sidewalls. The hot air in the plume is more buoyant than the original air in the room, and so the plume establishes a layer of hot air in the upper part of the room. Initially, this layer is very thin and close to the ceiling. However, the con-tinuing plume entrains hot air from this layer and arrives at the top of the space even hotter. The plume thus spreads out above the first layer of hot air, displacing the latter down-wards. In time, a considerable region of hot air is formed in the upper part of the space. This region is separated from the original cold air in the room by a density step called *the first*

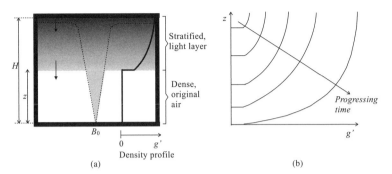

Figure 3.10. Evolution of the temperature profile in a room with no opening heated by a localised source. (Adapted from Worster & Huppert 1983.)

front, which is in fact the first thin layer of hot air from the plume that reaches the ceiling. This first front descends with time (Fig. 3.10), and for a room of height H and uniform cross-sectional area S, its dimensionless position at time t, $z(t)/H$, as measured from the level of the source may be given as a function of the room height H, the room cross-sectional area S, the initial buoyancy flux of the plume B_0 and the plume entrainment constant ε as (Baines & Turner 1969)

$$\frac{z(t)}{H} = \left[1 + \frac{1}{5}\left(\frac{18}{5}\right)^{1/3} 4\pi^{2/3} \varepsilon^{4/3} H^{2/3} S^{-1} B_0^{1/3} t\right]^{-3/2}. \quad (3.2.2\text{-i})$$

The evolution of temperature (and density) in the region above the first front (the buoyant region) is a function of the plume radius b, its vertical velocity u and the cross-sectional area S of the room (Worster & Huppert 1983):

$$\frac{\partial}{\partial t}\left(g\frac{\Delta\rho}{\rho_0}\right) = \frac{\pi b^2 u}{S}\frac{\partial}{\partial z}\left(g\frac{\Delta\rho}{\rho_0}\right). \quad (3.2.2\text{-ii})$$

The numerical results of Eqs. (3.2.2-i) and (3.2.2-ii) are sketched in Fig. 3.10b, based on the work by Germeles (1975) and Worster and Huppert (1983). The picture illustrates the characteristic 'filling box' process, that is, the filling of the room by warm fluid through the descending first front.

A connection can be readily made between the picture of flow in Fig. 3.10 and an accumulation of smoke or heat in an actual space whose doors and windows are closed. In such a space, the speed of descent of the smoke/heat layer depends on the size of the room and the strength of the source (e.g., the size of a fire, the power of a heat-emitting electrical device). The concentration of the smoke/heat progressively increases at a rate relative to the rate at which the room is filled. The design/control implications of this depend on the intended operation of the space. If the intention is to build up temperature to attain thermal comfort in winter, then subdividing the space into smaller, shorter spaces and/or using a stronger heater will allow the filling-box process to develop more quickly, enabling thermal comfort to be achieved more readily. On the other hand, if the desire is to prevent a rapid accumulation of smoke in the case of a fire or a build-up of gas in the case of a gas leak, then a taller, wider space will be more appropriate (Fig. 3.11). Note that, in the latter case, providing sufficient ventilation through vents made in the envelope of the space may prevent the pocket of smoke/gas from reaching the lower occupied zone altogether. Doing this, however, will eliminate the filling box process towards the end of the flow evolution, making the system evolve in a different manner. This is the situation that we look at in the next section.

(a)   (b)

Figure 3.11. A taller, wider space takes longer to heat but is safer in the
event of a fire or a gas leak.

3.2.3. Ventilated enclosure

If the space in Fig. 3.10 contains some openings connecting
its interior to the ambient environment, then there obviously
will be some ventilation. In this case, if the openings are made
in the envelope at high and low levels, as at the Centre for
Mathematical Sciences in Cambridge shown in Fig. 3.12, the
general form of ventilation that develops in the space will
follow the diagram in Fig. 3.13 (Linden, Lane-Serff & Smeed
1990). Basically, the plume rises from the base of the space
to the top in a manner similar to that observed in the room
with no opening discussed in Section 3.2.2. However, with the
room now containing openings, the warm layer, on becoming established in the upper part of the space, drives outflow
through the upper opening, because the hydrostatic pressure
difference between the top and bottom of this layer is smaller
than that between the same heights in the denser environment outside (see Section 2.2). Simultaneously, the buoyancy

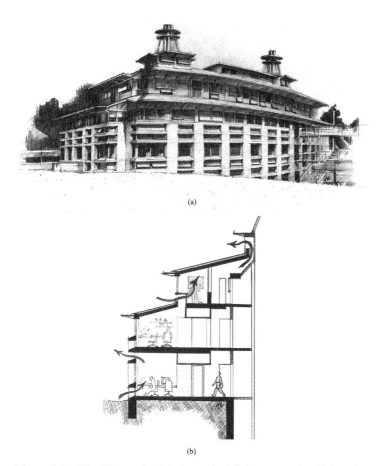

(a)

(b)

Figure 3.12. The Centre for Mathematical Sciences at the University of Cambridge uses a combination of high-level windows and low-level dampers to facilitate buoyancy-driven displacement ventilation through each office space.

provided by the plume draws ambient air into the room through the lower opening. Part of this ambient air is then entrained and carried by the plume to the upper layer. Initially, the warm upper layer deepens downwards, causing an

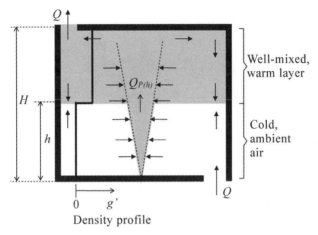

Figure 3.13. Steady displacement flow of a room with openings at two
levels heated by a localised source. (Following Linden, Lane-Serff &
Smeed 1990.)

interface between this layer and a layer of cold ambient air in
the lower zone to descend, while the mean temperature of the
upper layer increases. At this early time, convective mixing
is observed across the interface between the two layers. This
mixing, however, becomes less vigorous with time as the speed
of descent of the upper layer decreases and the density contrast between the upper and lower zones increases. Eventually, a steady state is reached (Fig. 3.13), at which the air in the
upper layer becomes well-mixed at a temperature above that
of the ambient environment, and the interface is arrested at a
height h such that the volume flux supplied by the plume at this
height is in balance with the outflow volume flux through the
upper opening. This latter volume flux, in turn, matches the
inflow volume flux through the lower opening. At equilibrium,

warm air still rises within the plume, but outside the plume at the level of the interface the vertical velocity becomes zero, although there is still a horizontal component of velocity towards the plume, thanks to the plume entrainment. Thus there is no convective mixing between the upper and lower layers across the interface at steady state, and the advection of air from the lower to upper layer is accomplished through the plume alone. We note here that, at the interface, thermal diffusion also takes place, causing the interface to become blurred. However, the influence of thermal diffusion on the overall flow is minimal when convection is strong, and may be neglected in leading order analysis. Adopting this simpli-fication, an expression for the net volume flux Q through the openings at steady state may be written in terms of the reduced gravity $g' = g\alpha\Delta T$ associated with the temperature excess ΔT above the exterior of the sharply defined upper layer, namely

$$Q = A^*[g'(H - h)]^{1/2}, \qquad (3.2.3\text{-i})$$

where A^* is the effective opening area given by Eq. (2.4-ii), and H is the vertical distance between the lower and upper openings. The term $[g'(H - h)]^{1/2}$ describes the speed of the flow. For a localised source of heat flux H_P, the temperature excess in the upper layer may be found from the conservation of thermal energy (Eq. (2.1.2-i)). This gives

$$\Delta T = \frac{H_P}{\rho C_P Q}, \qquad (3.2.3\text{-ii})$$

where ρ and C_P are the density and specific heat capacity of air, respectively. The volume flux supplied by the plume at the height of the interface, $Q_{P(h)}$, is given by Eq. (3.2.1-xiii), with $z = h$. Equating this equation with Eq. (3.2.3-i) gives the dimensionless height $\xi = h/H$ of the interface at steady state in the form

$$\frac{A^*}{H^2} = \lambda^{\frac{3}{2}} \left(\frac{\xi^5}{1 - \xi} \right)^{1/2}, \qquad (3.2.3\text{-iii})$$

where $\lambda = (6\varepsilon/5)(9\varepsilon/10)^{1/3}\pi^{2/3}$ has a typical value of 0.16. Eq. (3.2.3-iii) is, in fact, the same as the expression derived by Thomas, Hinkley, Theobald and Simms in 1963 from their study of the ventilation of a Boussinesq fire plume. This expression shows that, counter-intuitively, the height of the interface at steady state is independent of the source strength H_P but governed by the plume entrainment beha-viour (described by λ) and the geometry of the space, namely the effective opening area A^* and the vertical separation between the openings H. However, unlike the interface height, the temperature in the upper layer and the flow rate at steady state increase with the strength of the source: the former is proportional to the two-thirds power of the buoy-ancy flux (Eq. (3.2.1-xi)) while the latter is proportional to the one-thirds power of the buoyancy flux (Eq. (3.2.1-xiii)).

It should be noted that the general picture of flow in Fig. 3.13 also applies to the case where the plume does not rise from the middle of the space but in a corner or next to a wall. In this case, the shape of the plume does not follow the axisymmetric cone shown in Fig. 3.9, but is cut off. Therefore, its entrainment is reduced (smaller λ), causing the height ξ of the interface at steady state to elevate.

The picture of flow described in the preceding text goes some way to explain one curious phenomenon frequently encountered in real life: that turning up a localised heater sometimes does not lead to a perceivably warmer environment. This problem may be outlined thus. When a room operates in a cold season or contains a small heat load, its indoor temperature may fall below a comfort zone. To alleviate discomfort it is customary to fit the room with a localised heat source of some kind, for example, a wall-mounted radiator. In this situation, if the room is relatively small and conventionally occupied, the heat output from the radiator (which is of order 10^3 W each) is likely to dominate the heat output from the occupants (which is of order 10^2 W per person) or that of lighting and other small electrical equipment (which is of order 10–10^2 W each). Thus the overall behaviour of flow in the space is likely to be controlled by the behaviour of the thermal plume generated by the radiator, so that the room becomes stratified essentially as in Fig. 3.13: the upper part of the room is supplied primarily by warm air from the radiator while the lower part contains air at the ambient temperature. (In fact, due to the presence of multiple sources of buoyancy in the space, the actual temperature structure will be more complex than that described, although to leading order the picture given is sufficient for the present discussion; we examine the complex temperature structure arising from multiple buoyancy sources in Section 3.2.5.) If the height of the interface between the warm and cold zones lies above the occupied zone (greater than 2 m from the floor level, say), the occupants will feel chilly. Their reaction in this situation will usually be to turn up the radiator, with the hope of making the air around them warmer. However, our earlier discussion suggests that doing so will only increase the temperature in the upper part of the room, without affecting the temperature in the lower occupied zone in the long term. (Note, though, that as the radiator is turned up, the interface may descend *temporarily*, due to a rise in the plume volume flux following the increase in the heat input; see Section 3.2.4. Also, if the occupants are sufficiently close to the radiator, they may experience a higher temperature due to increased radiation. This radiation, nonetheless, accounts for only a small percentage of the heat given off by the radiator – a much higher percentage of the heat is distributed by means of convection – and can be easily obstructed by a piece of furniture.)

To bring about thermal comfort in the lower zone, the system will need to be adjusted according to the flow principles outlined earlier. Equation (3.2.3-iii) shows that the interface height at steady state is sensitive to the effective size of the openings; therefore, to bring down the warm upper layer to the occupied zone, windows and doors must be closed or their gaps/cracks sealed up sufficiently (with weather strips, for example) to reduce the ventilation flow through the room. Obviously, doing so will reduce the amount of fresh air entering the room, potentially leading to insufficient ventilation. However, experience in the field has shown that infiltration through small cracks and gaps in the walls, doors and windows usually supplies more than enough fresh air to allay any concern about poor indoor air quality in small spaces of typical construction, provided that the spaces are not too densely occupied.

The flow principles outlined in this section are also relevant to air-conditioned spaces. In such a space, the air-conditioning unit supplies a down-welling, negatively buoyant cold plume, leading to a layer of cold air being formed at the base of the space. If the room is sufficiently leaky, this cold layer may be shallow enough compared with the depth of the occupied zone so that the occupants have a sensation of cold feet, owing to an excessive temperature contrast between their upper body zone and their lower leg zone. This can lead not only to discomfort but also to ill health, depending on the severity of the temperature contrast and the period over which the occupants are exposed to the undesirable conditions. To prevent this problem, the windows and doors will have to be closed/sealed properly to allow the cold layer to deepen sufficiently to match the depth of the occupied zone. The temperature within the cold zone may then be controlled by adjusting the flux of cooling from the air-conditioning unit.

3.2.4. Transient responses

The steady-state flow picture outlined in Section 3.2.3 may be inadequate in certain circumstances to allow a system of ventilation to be controlled effectively: a heater may be turned up or down or people may leave or enter the space while the ventilation is still developing, causing the system to respond transiently. This problem was studied by Bower, Caulfield,

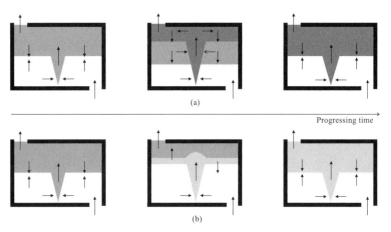

Figure 3.14. Transient evolution of the temperature structure of a room heated by a localised source, when the heat flux is increased instantaneously (a) and when the heat flux is reduced instantaneously (b). (Following Bower, Caulfield, Fitzgerald & Woods 2008.)

Fitzgerald and Woods in 2008. They showed theoretically and experimentally that when the strength of the heat load in the space is varied, two basic transient flow regimes can develop, depending on whether the heat load is increased or reduced from its original value.

Assuming that the change in the heat load is instantaneous (as may be the case when adjusting the output of an electric heater, for instance), if the heat load is *increased*, the continuing plume becomes more buoyant than the upper layer. Consequently, the plume rises to the top of the space and forms a new, warmer layer above the original layer (Fig. 3.14a). This new hotter layer deepens with time to fill the upper part of the room and, because the layer is supplied by the plume that has entrained fluid from the lower original layer, it also causes the original layer to deplete. Initially, the original lower interface descends in response to the increased plume volume flux caused by the increase in the heat load (Eqs. (3.2.1-xiii) and (3.2.1-xiv)). This lower interface, however, gradually rises back over time to match the geometry of the room (Eq. (3.2.3-iii)). Finally, a new equilibrium is reached at which the room becomes stratified into two layers, with the new filling layer replacing completely the original buoyant layer. The temperature in this new upper layer is greater than that of the original layer, owing to the increased heat load (Eq. (3.2.3-ii)), but the interface dividing this layer from the layer of ambient air underneath remains at the same height as at the previous equilibrium, because the geometry of the room is unchanged (Eq. (3.2.3-iii)). The flow rate, flow speed and interior temperature at the new steady state may be calculated using a combination of Eq. (3.2.1-xiii) and Eqs. (3.2.3-i)–(3.2.3-iii). Figure 3.15a depicts the described evolution of the interface positions. This picture of flow can be quantified by considering the mass balance of the system in relation to the volume flux driven by the total buoyancy associated with the two buoyant layers, taking into account the contribution from the enhanced plume and the effect of reduced plume buoyancy as the plume rises through a stratified environment into the upper layer. The timescale of evolution of the system depends chiefly on the magnitude of change in the heat load. If the increase in the heat load is large, the timescale of evolution is controlled by the new heat load, so that the original buoyant layer has a secondary influence on the flow. As a result, the transient readjustment to the new steady state

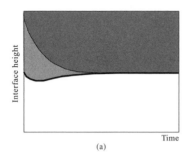 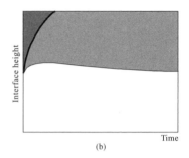

(a)   (b)

Figure 3.15. Transient evolution of the temperature structure in a room heated by a localised source, when the heat flux is increased instantaneously (a) and when the heat flux is reduced instantaneously (b). In each figure, the original interface is indicated by the thicker line, and the new interface by the thinner line. In (a), the original upper layer is shown in lighter grey and the new hotter layer in darker grey. In (b), the original upper layer is shown in darker grey, and the new less buoyant intermediate layer in lighter grey. (Based on calculations by Bower, Caulfield, Fitzgerald & Woods 2008.)

occurs over the timescale required for the enhanced plume to fill the room (i.e., the 'filling box' timescale, t_F, defined as $t_F = SH/Q_{(H)}$, where S is the cross-sectional area of the room, H is the height of the room and $Q_{(H)}$ is the volume flux of the enhanced plume at height H). With an existing buoyant layer in the space, the time required to readjust to a new steady state is longer than it would otherwise be if the space were completely empty initially (the case investigated by Kaye & Hunt 2004). This is because the presence of an existing buoyant layer enhances the outflow through the upper opening, delaying the approach of the filling layer.

The evolution of the system will be quite different if the heat load is *reduced* instantaneously from its original value

(Figs. 3.14b). In this case, the continuing plume is less buoy-ant than the original buoyant layer. Consequently, the plume only rises to the interface between the original buoyant layer and the layer of ambient air, forming a new intermediate layer of temperature between the two layers. Because the volume flux of the plume is reduced as a result of the reduction in the heat load (Eqs. (3.2.1-xiii) and (3.2.1-xiv)), the total depth of the two buoyant layers initially decreases. As the ventil-ation progresses, however, this depth gradually increases to match the geometry of the room (Eq. (3.2.3-iii)). Meanwhile, the intermediate layer deepens and the original upper layer depletes owing to a combination of the ventilation through the top opening and, in the case where the plume is of sufficient momentum, *penetrative convection* at the original (upper) interface. This penetrative convection occurs as a result of the relatively dense plume overshooting into the original upper layer and, on collapsing back into the intermediate layer, con-vecting down air from the original layer and mixing it with air in the intermediate layer. For a system in which the original buoyant layer is shallow and the reduction in the heat load is large, penetrative convection is insignificant, hardly affecting the overall flow evolution. In contrast, if the original buoyant layer is deep or, in particular, if the reduction in the heat load is small, penetrative convection has a considerable impact on the overall flow. Eventually, the system converges to a new steady state at which the original buoyant layer is completely depleted and the room becomes stratified into two layers. The new upper layer is cooler than the original one owing to the decrease in the heat load (Eq. (3.2.3-ii)). The interface, however, lies at its original position because the geometry of the room is unchanged (Eq. (3.2.3-iii)). The flow rate, flow speed and interior temperature at the new steady state can be estimated using a combination of Eq. (3.2.1-xiii) and Eqs. (3.2.3-i)–(3.2.3-iii). The evolution of the interface positions in this transient flow regime is shown in Fig. 3.15b. This picture of flow may be quantified by considering the mass balance of the system in relation to the ventilation flow driven by the total buoyancy associated with the two buoyant layers, with the volume flux of penetrative convection at the upper interface and the volume flux of the reduced plume at the lower interface taken into account. The evolution of the temperature of the intermediate layer is influenced by both the temperature of the arriving weakened plume and, in the case in which there is noticeable penetrative convection, the temperature of the air entrained from the hotter original layer. The time taken for the system to converge to the new steady state is observed to be related to but somewhat shorter than the timescale of draining associated with the original layer. With the heat load reduced but still present, the original layer drains faster than it otherwise would if the source were removed completely (i.e., $B_0 = 0$). This is because the added buoyancy from the intermediate layer increases the ventilation flow.

The practical implications of the aforementioned transient flow pictures may be appreciated by considering a simple situation outlined originally by Bower, Caulfield, Fitzgerald and Woods (2008). Suppose a room is heated by a localised source as in Fig. 3.13. A control system may be set up for the space so that, within a certain range of exterior temperatures, the interface between the warm upper layer and the lower layer of ambient air is kept well within the occupied zone so as to allow the occupants to benefit from the warm upper layer. If the exterior temperature rises beyond a prespecified limit, the heating is decreased. This will trigger the readjustment of the interior temperature structure according to the regime in Figs. 3.14b and 3.15b. The continuing plume from the weakened source will form a cooler intermediate layer below the original interface. Therefore, the occupants in the lower occupied zone will benefit from the reduction in the magnitude of heating straightaway.

Conversely, if the exterior temperature falls below a prescribed value, the heating is turned up. In this case, the room will readjust following the regime in Figs. 3.14a and 3.15a. The hotter plume from the heater will rise to the top of the space before gradually propagating down while eroding the original upper layer. The time taken for the new, hotter upper layer to descend to the level of the original interface (and therefore to the occupied zone) will depend greatly on the size of the space and the ventilation rate. If either or both of these are large, it will be some time before the occupants in the lower zone can enjoy the increased comfort associated with the enhanced heating. Therefore, a supplementary system of heating may be required to maintain thermal comfort during the readjustment period. This may be provided locally in the form of a heated seat or pre-heated air supplied through a floor diffuser, for example.

A connection can also be made between the transient flow pictures described above and an air-conditioned space.

Assume that the ventilation system is made separate from the cooling system to ensure that an appropriate amount of fresh air is always achieved regardless of the magnitude of cooling, and that the mass flux of air from the cooling system is zero or negligible compared with the mass flux driven by the ventilation (as may be the case in a room equipped with a localised passive chilled beam, for instance). If cold air is supplied from a high level, for example, from the ceiling, the response of the system will essentially be an upside-down version of Figs. 3.14 and 3.15. A control system can be designed so that a pool of cold air completely occupies the lower occupied zone while the zone above is filled with ambient air. If the cooling flux is increased from its original value, after an increase in the exterior air temperature for instance, the colder continuing plume will descend straight to the base of the room before deepening back to the level of the original interface. This will allow the occupants to enjoy the increased comfort immediately. In contrast, if the cooling flux is reduced from its original value after a decrease in the exterior air temperature, the less cold continuing plume will form a layer above the original cold layer. The occupants will not benefit from this new, less cold layer until the original layer has been depleted sufficiently by a combination of ventilation through the lower opening and erosion caused by penetrative convection. Depending on the time taken for the system to readjust, a complementary system of heating may be necessary initially to keep the occupants comfortable.

Care must be taken, however, in applying the foregoing principles to an air-conditioned space in which the mass input of cold air from the air-conditioning system is considerable compared with the mass flux driven by the ventilation (such as a room fitted with a conventional all-air air-conditioning system). In this case, if the mass input from the air-conditioning unit is proportional to the cooling flux (as is often the case), the layer of cold air may deepen quickly as the cooling flux is increased, leading to a blockage of ventilation. Detailed analysis of this situation is complex and beyond the scope of this book; interested readers are advised to consult Woods, Caulfield and Phillip (2003) instead.

Before we end this section, it will be useful to talk a little more about penetrative convection mentioned in various places earlier; the phenomenon will be encountered again in other flow problems featured in this book. As stated previously, penetrative convection occurs when a plume, possessing sufficient momentum, rises (or falls, in the case of a negatively buoyant plume) through a layer of fluid and impinges on another layer of fluid of different density, before collapsing back, bringing with it the fluid from the impinged layer. This form of penetrative convection has been studied, among others, by Linden (1973); Baines (1975); McDougall (1981); Kumagai (1984); Baines, Turner and Campbell (1990); Bloomfield and Kerr (2000); Lin and Linden (2005) and Bower, Caulfield, Fitzgerald and Woods (2008); and detailed discussions on how to model its mechanism may be obtained from these studies. Baines, Turner and Campbell (1990), for example, analysed experimental data from many authors and derived an expression for estimating the height of the penetrating plume as a function of its momentum flux, its buoyancy flux and an empirically determined constant (whose value was later revised by Bloomfield & Kerr in 2000). This expression enables the determination of whether or not the plume rises beyond the upper interface and therefore actually has potential to entrain fluid from the upper layer. Linden (1973), Baines (1975) and Kumagai (1984) modelled the flux of entrained fluid in terms of the interfacial Froude number. This dimensionless number measures the ratio of the inertia of the arriving plume to the relative density jump at the original layer. Bower, Caulfield, Fitzgerald & Woods (2008) modelled the flux of entrainment more simply as linearly proportional to the volume flux of the plume. The physics and mathematics of penetrative convection will be invoked again when the phenomenon is revisited in Sections 3.2.5, 3.3.2, and 3.4.

3.2.5. Multiple localised sources

While a room may contain a single source of buoyancy or one which is dominant over the others (as in the example given at the end of Section 3.2.3), it is also possible for a room to contain multiple sources of buoyancy of comparable strength, as in, for example, the office shown in Fig. 3.16. In this case, the form of internal stratification that develops depends chiefly on the relative strength of the sources (Cooper & Linden 1996; Linden & Cooper 1996). To examine flows in this situation it is convenient to begin with the assumption that the sources are located sufficiently far apart from one another, so that their plumes do not collide, and that the plume theory described in

Figure 3.16. In a typical office where the occupants carry out similar activities, each work unit may be regarded as a localised source emitting a comparable amount of heat.

Section 3.2.1 directly applies. This is an analogue of a real-life space in which there are a number of occupants working a reasonable distance away from each other. We will address the situation where the plumes are closely positioned and collide at the end of the section.

If there are n localised sources of *identical* strength in the space, n equally buoyant plumes are produced, leading to two-layered stratification at steady state, not dissimilar to that achieved in a room containing a single localised source (cf. Fig. 3.17 and Fig. 3.13): plumes of the same density form the same buoyant layer. At steady state, an upper layer of

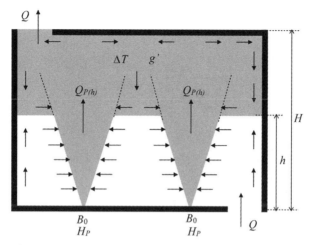

Figure 3.17. Steady displacement flow in a room with openings at two levels heated by two localised sources of equal strength. The diagram can be generalised for n number of sources of equal strength, with the interface descending as the number of sources increases. (After Cooper & Linden 1996.)

reduced gravity g' is separated from a lower layer of ambient air at height h by a density interface (which, as we may recall from our previous discussion, is somewhat blurred in reality but may be regarded as sharp in simple analysis). It is this reduced gravity associated with the warm, buoyant layer that drives the ventilation. At steady state, the net volume flux Q through the building matches the total volume flux $nQ_{P(h)}$ supplied to the upper layer by all the plumes combined by virtue of mass conservation,

$$Q = A^* [g'(H-h)]^{1/2} = nQ_{P(h)} = n\lambda B_0^{1/3} h^{5/3}, \qquad (3.2.5\text{-i})$$

where $nQ_{P(h)}$ is given according to Eq. (3.2.1-xiii), and the value of g' can be obtained from the conservation of thermal energy in the upper layer. If each plume source has a heat flux H_P, we have

$$g' = g\alpha \Delta T = g\alpha \frac{n H_P}{\rho C_P Q}. \qquad (3.2.5\text{-ii})$$

Solving Eq. (3.2.5-i) will lead immediately to an expression for the dimensionless height of the interface at steady state, $\xi = h/H$, namely

$$\frac{1}{n}\frac{A^*}{H^2} = \lambda^{3/2}\left(\frac{\xi^5}{1-\xi}\right)^{1/2} \qquad (3.2.5\text{-iii})$$

which is, in fact, an algebraic extension of Eq. (3.2.3-iii). Eq. (3.2.5-iii) indicates that, as with the case of the room with one buoyancy source, the height of the interface in the room with multiple buoyancy sources of equal strength depends on the entrainment of the plumes and the room geometry, namely the effective vent size A^* and the vertical separation between the vents H, but is unaffected by the total buoyancy flux of the sources. The only departure from the case of one source is that, with multiple sources, the height of the interface is also dependent on the number of sources, n. Fig. 3.18, plotted using Eqs. (3.2.5-i)–(3.2.5-iii), clarifies this statement graphically. It can be seen that when there are more sources in the space, the upper layer becomes deeper (Fig. 3.18a). For a fixed total heat flux, this leads to a faster flow (Fig. 3.18b), thanks to an increased buoyancy head. Moreover, it leads to a colder upper layer (Fig. 3.18c), owing to the distribution of heat over a deeper buoyant layer.

The height of the interface, it may also be observed, is very sensitive to the number of sources, descending rapidly as

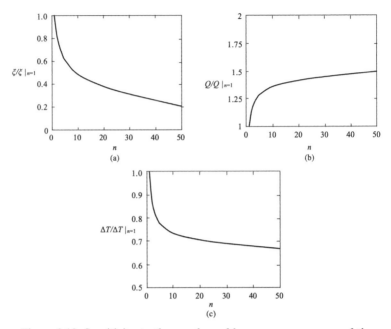

Figure 3.18. Sensitivity to the number of buoyancy sources, n, of the interface height ξ (a), the flow rate Q (b) and the temperature in the upper layer ΔT (c) in a room containing multiple localised sources of buoyancy of identical strength. The interface height, flow rate and temperature are shown as relative to their values when there is only one buoyancy source in the room ($n = 1$). In these plots, the total heat flux is fixed and typical values of entrainment constant $\varepsilon = 0.1$ and $A^*/H^2 = 0.0375$ are taken. (Based on the models by Cooper & Linden 1996 and Turner 1973.)

the number of sources increases (Fig. 3.18a): distributing the buoyancy flux from a single source into 10 equal sources will just about halve the depth of the lower ambient zone. This has direct implications for the ventilation of fire and light contam-inant gases in particular, as it shows that the lower smoke/gas-free zone depletes quickly as the number of fires/gas leaks

increases. As the number of sources becomes very large, the
interface descends very close to the floor, leading to the whole
room essentially being filled with smoke/gas.

For the case where the strengths of the sources are
unequal, the resultant flow structure becomes complex. This
situation represents what often happens in real life: it is common for a number of sources of different magnitudes, such
as a person and electrical devices, to be present simultaneously in the same space (Fig. 3.19). The plumes generated
in this case are unequally buoyant, and so the hottest, most
buoyant plume forms a layer immediately underneath the
ceiling, while the others form subsequent layers in order of
descending buoyancy. The number of stratified layers follows
the number of sources that are different in strength.

It is convenient to gain an entry to the problem of multiple sources of unequal strength by considering the simple
case of just two buoyancy sources of unequal strength. Suppose that these sources are of flux B_1 and B_2, and that $B_1 \leq B_2$
(Fig. 3.20). The room becomes stratified at steady state into
three layers. The uppermost layer is formed by the most buoyant, hottest plume from source B_2, the intermediate layer is
formed by the less buoyant, weaker plume from source B_1,
and the bottommost layer is formed by non-buoyant ambient air. The interfaces between the three layers are stable at
steady state, and air is transferred across the layers through
the plumes only. Therefore, the volume flow into and out of
the space may be related to the plumes' volume fluxes and the
reduced gravity associated with each buoyant layer. Let us use
a double subscript system to describe the volume flux of each

Figure 3.19. In spaces such as the study room depicted here, a number of localised sources, such as a person, a heater, a computer and a lamp, may be present simultaneously, each having a different strength.

plume at each level, with the first term denoting the source and the second term the level of the interface; for example, Q_{21} refers to the volume flux of the plume from source B_2 at the height of the lower interface. If source B_1 is sufficiently weak, it will not overshoot the upper interface. In this case, penetrative convection, denoted by Q_{PC} in Fig. 3.20, can be ignored (we will discuss the effect of penetrative convection

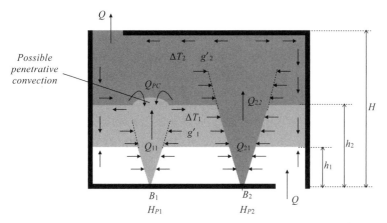

Figure 3.20. Diagram of steady displacement flow in a room with openings at two levels heated by two localised sources of unequal strength. (Following Linden & Cooper 1996.)

later). It then follows from the conservation of mass that the net volume flux Q of the system is matched by the volume flux Q_{22} of the stronger plume at the level of the upper interface, and that this volume flux Q_{22}, in turn, equals the sum of the volume flux Q_{11} entrained from the intermediate layer by the stronger plume and the volume flux Q_{21} of the stronger plume at the level of the lower interface:

$$Q = Q_{22} = Q_{11} + Q_{21}. \qquad (3.2.5\text{-iv})$$

Application of Bernoulli's theorem quickly gives that the net volume flux Q is driven by the total reduced gravity acting over the two buoyant layers, namely

$$Q = A^*[g_2'(H - h_2) + g_1'(h_2 - h_1)]^{1/2}, \qquad (3.2.5\text{-v})$$

where g_1' and g_2' are the reduced gravity associated with the intermediate and topmost layer, respectively, and h_1 and

h_2 are the level of the lower and upper interface as measured from the bottom opening, respectively. The intermediate layer is heated and supplied by the weaker plume alone, and so the conservation of thermal energy gives g'_1 as a function of the heat flux H_{P1} of the weaker plume and the volume flux Q_{11} to the intermediate layer,

$$g'_1 = g\alpha\Delta T_1 = \frac{g\alpha H_{P1}}{\rho C_P Q_{11}}, \qquad (3.2.5\text{-vi})$$

with ΔT_1 being the temperature excess in the intermediate layer above the ambient air. The uppermost layer, on the other hand, is heated by a combination of the stronger plume and heat entrained by this plume as it passes through the intermediate layer. Taking into account the volume flow Q_{22} through the uppermost layer (which is the same as the net flow Q), we obtain from the conservation of thermal energy

$$g'_2 = g\alpha\Delta T_2 = \frac{g\alpha(H_{P1} + H_{P2})}{\rho C_P Q_{22}} = \frac{g\alpha(H_{P1} + H_{P2})}{\rho C_P Q},$$
$$(3.2.5\text{-vii})$$

where ΔT_2 is the temperature excess in the uppermost layer, and H_{P2} is the heat flux of the stronger source. Solving the mass and heat balance equations above and applying the plume theory leads directly to the expression of the heights of the interfaces, h_1 and h_2, in the form

$$\frac{A^*}{H^2} = \lambda^{3/2}\frac{(1+\psi^{1/3})^{3/2}}{(1+\psi)^{1/2}}\left[\frac{\left(\frac{h_1}{H}\right)^5}{1 - \frac{h_1}{H} - \frac{(1-\psi^{2/3})}{(1+\psi)}f(\psi)\frac{h_1}{H}}\right]^{1/2};$$
$$f(\psi) = \frac{h_2}{h_1} - 1, \qquad\qquad (3.2.5\text{-viii})$$

where $\psi \equiv B_1/B_2 \leq 1$ denotes the ratio of the buoyancy fluxes of the two sources, and $f(\psi)$ is a function of the ratio of the source strengths.

Again, as with the cases of a single source and multiple sources of equal strength, the heights of the interfaces in the room with multiple sources of unequal strength are independent of the total buoyancy flux, but dependent on the geometry of the room, A^* and H. However, with the sources now being different in strength, the interface heights are also influenced by the *ratio* ψ of the source strengths.

Clearly, to identify the position of each interface, the value of $f(\psi)$, which describes the ratio of the interface heights, must first be obtained. The explicit expression for the interface height ratio h_2/h_1 is quite involved and given in full in Cooper and Linden (1996). Here, we merely note the principle that to identify the interface height ratio, the behaviour of the stronger plume as it crosses the lower interface and its subsequent motion in the intermediate layer need to be considered. Detailed modelling is given by Linden and Cooper (1996), but for the present analysis it is adequate just to recall a principle in the plume theory (Section 3.2.1) that states that the buoyancy flux of a plume is conserved so long as the plume rises through an environment of uniform density, but is reduced as the plume crosses from an environment of one density to another environment of a lower density. This is precisely what happens to the stronger plume in Fig. 3.20 as it travels through the intermediate layer into the uppermost layer. The lower part of this plume rising through the ambient air can be treated as an unforced plume originating from a source of zero volume flux and zero momentum flux, while its upper part rising through the intermediate layer can be treated as originating from a source of finite momentum flux and finite volume flux located at the lower interface. It can be shown that the buoyancy flux of the stronger plume within the intermediate layer, B'_2, has the value

$$B'_2 = B_2 - g'_1 Q_{21},\qquad\qquad\text{(3.2.5-ix)}$$

and that the ratio of B'_2 to its original value, B_2, is a function of the ratio of the source strengths, namely

$$\frac{B'_2}{B_2} = (1 - \psi^{2/3}).\qquad\qquad\text{(3.2.5-x)}$$

Cooper and Linden (1996) investigated how the temperature profile in the space changes as the ratio ψ of the source strengths is varied. For the case in which penetrative convection at the upper interface is ignored, the results are shown as the dotted lines in Fig. 3.21a. The plots are obtained based on a room of certain geometry, but the principles captured by them carry through to other cases of different geometries. It can be seen that, for $\psi = 0$, corresponding to the case of one buoyancy source, only a single interface forms. In this case, Eq. (3.2.5-viii) reduces to Eq. (3.2.3-iii), recovering the result of a single plume given by Linden, Lane-Serff and Smeed (1990). For $1 > \psi > 0$, corresponding to two buoyancy sources of unequal strength, the interface splits into two, with the lower interface descending below the original interface and the upper interface established above the original interface. As the relative strength of the weaker source increases,

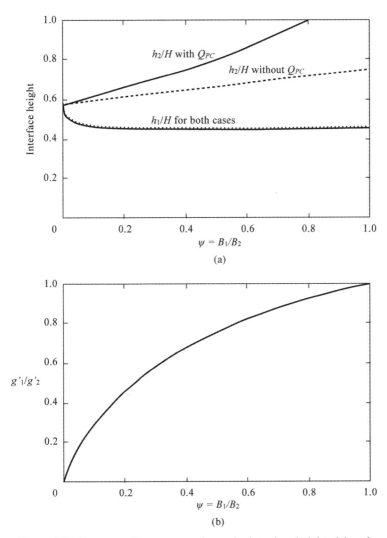

Figure 3.21. Impacts of buoyancy ratio on the interface heights (a) and the relative buoyancy of the upper and lower layers (b). (Based on models by Cooper & Linden 1996 and Linden & Cooper 1996.)

$\psi \to 1$, the upper interface ascends and the lower interface descends. Meanwhile, the buoyancy of the intermediate
layer relative to that of the uppermost layer, g'_1/g'_2, increases
(Fig. 3.21b). This relative buoyancy is controlled by the ratio
of the source strengths only, following the relation

$$\frac{g'_1}{g'_2} = \frac{(\psi + \psi^{2/3})}{(1 + \psi)}. \tag{3.2.5-xi}$$

Eventually, when the strengths of the two sources are equal,
$\psi = 1$, the buoyancy of the two buoyant layers become
identical, $g'_1 = g'_2$ and the interface between the intermediate
and uppermost layers disappears. At this point, the height of
the remaining lower interface can be calculated using Eq.
(3.2.5-iii), with $n = 2$. Note, the lower interface descends
quickly only up to $\psi \approx 0.2$, beyond which it remains quite
steady, even when ψ approaches unity. This indicates that
the depth of the contaminant-free lower zone remains quite
constant regardless of the distribution of buoyancy between
the sources, and that, in practice, this depth can be estimated
using the simple relation (3.2.5-iii).

As mentioned earlier, the temperature structure depicted by the dotted lines in Fig. 3.21a does not take into account
penetrative convection at the upper interface (denoted by
Q_{PC} in Fig. 3.20). This penetrative convection is, in essence,
the same as the one observed in Section 3.2.4 in the case of
a single localised source. It develops when the weaker plume
is sufficiently strong (i.e., when the value of ψ is sufficiently
large) that it possesses enough momentum to overshoot the
upper interface. Because the penetrative convection entrains

air from the uppermost layer down to the intermediate layer, it modifies the mass balance of the system, affecting the height of the upper interface. As shown by the upper solid line in Fig. 3.21, with penetrative convection taken into account, the upper interface ascends more quickly as the source strength ratio increases, $\psi \to 1$, reflecting the increased importance of entrainment of the upper layer air as the inertia of the weaker plume increases. For low values of ψ, the impact of penetrative convection is small and can be ignored. However, for large values of ψ, its impact is significant and must be taken into account if the height of the upper interface is to be estimated accurately. Approaches for modelling penetrative convection have been proposed by many researchers, and we have summarised them at the end of Section 3.2.4. For the case of three-layered stratification shown in Fig. 3.20, the volume flux convected down by the penetrative convection may be given rather conveniently in terms of the interfacial Froude number, Fr, and the volume flux of the weaker plume at the height of the upper interface, Q_{12}, as (Cooper & Linden 1996; Baines 1975; Kumagai 1984),

$$Q_{PC} = \frac{Fr^3}{1 + 3.1 Fr^2 + 1.8 Fr^3} Q_{12}, \qquad \text{(3.2.5-xii)}$$

where the Froude number is defined as the ratio of the plume inertia, represented by its vertical velocity u_{12}, to the drive due to the density step $(g'_2 - g'_1)$ at the upper interface; that is,

$$Fr = \frac{u_{12}}{[r_{12}(g'_2 - g'_1)]^{1/2}}, \qquad \text{(3.2.5-xiii)}$$

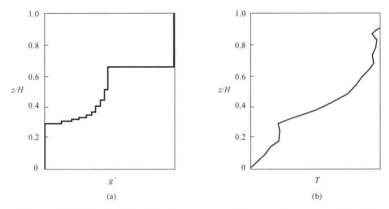

Figure 3.22. (a) Stratification in a room with openings at two levels heated by $n = 10$ localised sources of unequal strength (as calculated by Linden & Cooper 1996), and (b) a temperature profile taken from a real-life office containing a number of localised sources (based on Arnold 2004). Note that the temperature profiles in the two pictures do not match exactly as they do not represent the same environment; nonetheless, they both show similarly smooth stratification typical in spaces heated by multiple localised sources of unequal strength.

with r_{12} being the radius of the weaker plume at the upper interface.

The principles obtained from the case of two sources of unequal strength can be extended to the case of n sources of unequal strength. If the total heat flux of these sources is fixed, and if the sources are distributed so that their strengths relative to one another are $B_1 < B_2 < B_3$ and so on until B_n, then $n + 1$ layered stratification develops. The interior temperature profile becomes increasingly smoothly stratified as the number of sources increases, resembling what is usually observed in actual buildings (e.g., Fig. 3.22; see, among others, Gorton & Sassi 1982; Jacobsen 1988; Cooper & Mak 1991 and Arnold

2004). This is because real-life buildings often contain sources of buoyancy of different magnitudes located at various positions within the interiors. In principle, solutions similar to Eq. (3.2.5-viii) can be found for the positions of all the interfaces when there are n sources in the space. However, the procedure involved in doing so becomes increasingly unwieldy as the number of sources increases, owing to the need to readjust the virtual origin of each plume every time it crosses a density interface to account for the reduction in its buoyancy flux. This complication can be conveniently ignored in many practical circumstances, however, for it is usually the height of the lowest interface that determines whether or not the pocket of warm air/smoke lies within the lower occupied zone; it turns out that even though the heights of the higher interfaces (i.e., from h_2 upwards) are sensitive to the ratios of the source strengths and the number of sources, the height of the lowest interface is not, provided that the number of sources is sufficiently large. See Fig. 3.23. In this figure, the relative height of the lowest and immediately upper interfaces, h_2/h_1, is plotted against the number of sources, when there are $(n-1)$ equal plumes of strength B_i and one strong plume of strength B_n, such that $\psi_i = B_i/B_n \ll 1$ where $i = 1, \ldots, n-1$. The solid curve represents the case where $\psi_i = 0.01$, and the dotted curve represents the case where $\psi_i = 0.5$. It can be seen that when $\psi_i = 0.01$ and there are 20 plumes in the space, so that the weak plumes together contribute about 20% of the total buoyancy, the ratio of the interface heights is $h_2/h_1 \approx 1.02$. When the number of sources and their buoyancy distribution are vastly varied to $n = 30$ and $\psi_i = 0.5$, so that the total flux of the weak

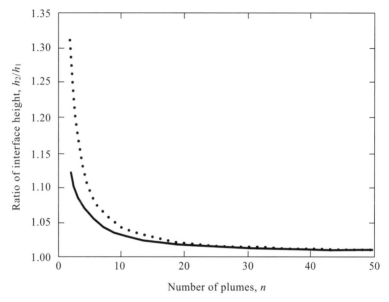

Figure 3.23. Ratio of interface heights, h_2/h_1, when there are $(n-1)$ equal plumes of strength B_i and one strong plume of strength B_n, such that $\psi_i = B_i/B_n \ll 1, i = 1, \ldots, n-1$. The solid curve is for $\psi_i = 0.01$ and the dotted curve is for $\psi_i = 0.5$. (Based on calculations by Linden & Cooper 1996.)

plumes is about 15 times that of the strong plume, the relative interface height hardly deviates, taking the value of $h_2/h_1 \approx 1.015$. The deviation of the interfaces becomes even less as the number of sources increases further. This shows that, as the number of sources increases, the interfaces approach each other so that the depth of the lower ambient zone becomes well represented by n equal sources, regardless of the distribution of buoyancy among them. Thus the simple expression (3.2.5-iii), which was given for n sources of equal strength, can also be used for estimating the depth of the lower zone when there are n sources of unequal strength. This relative insensitivity of the depth of the lower zone to the number of sources and their buoyancy distribution reflects the fact that the position of the interface is strongly controlled by the plume volume flux, which in turn is a strong function of the vertical distance from the source but a weak function of the buoyancy flux (Eq. (3.2.1-xiii)).

To control the depth of the lower zone, we must, therefore, control the geometry of the space rather than the distribution of the sources or their number. Eq. (3.2.5-viii) shows that the position of the lowest interface is sensitive to the effective size of the openings, A^*, and their vertical separation, H. However, openings sufficiently large or sufficiently far apart to lift the lowest interface to the desired level may be difficult to achieve in actuality, or even impossible, depending on architectural and engineering constraints among other things. For instance, a large opening or a tall ventilation chimney may not agree well with the overall aesthetics of the building or they may cause undue structural complications. Furthermore, because the effective opening area is essentially controlled by the size of the smaller opening (Fig. 2.5), increasing the sizes of openings at one location (e.g., windows at a high level) without increasing the sizes of those at others (e.g., doorways and dampers at lower levels) may not be able to lift the lowest interface sufficiently. Incorporating a soffit space to contain a pocket of smoke/gas could be a better solution in these circumstances.

A remark is due on what will happen if the sources are placed quite close to one another instead of far apart as in the cases examined in the preceding text. This situation is frequently encountered in real life: a printer may be placed beside a photocopier or two persons may sit right next to each other. In this case, the plumes, entraining ambient air between them, are naturally drawn together, and may collide and merge as they rise. The merging of the plumes affects the plume entrainment and therefore the height of the inter-face. Kaye (1998) showed that the merging process depends primarily on the plumes' relative strength and their separa-tion. For plumes of identical buoyancy, merging takes place about three to four source separations above the sources. For plumes of unequal strength, the weaker plume is drawn into the stronger plume, causing merging to take place sooner. In both cases, once the plumes have merged, the total buoyancy flux is conserved, but the volume flux in the combined plume is smaller than would be carried by the separate plumes. As a result, the height of the interface resulting from the merged plume is greater than that resulting from the separate plumes. However, closely placed plumes of infinite number will lead to the interface reaching the floor, and hence a well-mixed interior, as will separate plumes of infinite number (the latter is the asymptotic solution of Eq. (3.2.5-iii)). In Section 3.3 we examine this situation of a well-mixed room in the context of areal heating.

3.3. The distributed source

We will now look at the other end of the spectrum of source geometries, that is, a source of heating/cooling that is evenly

Figure 3.24. The International Digital Laboratory at the University of Warwick.

distributed throughout the space. This kind of source is an analogue of many buoyancy sources commonly found in modern buildings, including distributed occupants, a heated floor, a chilled ceiling and a surface through which heat is relatively easily lost, such as a glass roof. The International Digital Laboratory at the University of Warwick shown in Fig. 3.24

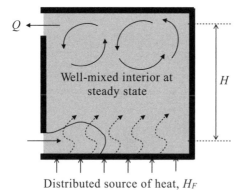

Distributed source of heat, H_F

Figure 3.25. Steady-state flow driven by a distributed source of heat located at the base of a space with openings at two levels. (After Gladstone & Woods 2001.)

is a real-life example of a naturally ventilated building that contains such buoyancy sources. Here, underfloor heating is used on its multiple floors both to provide comfort and to drive natural ventilation. The basic ventilation flows of this building and those of others subjected to comparable forms of heating/cooling may be modelled at small scale using a uniformly distributed array of heating elements or a hot plate, whose output is controlled electronically. Our analysis below will cover both steady-state flow regime and various forms of transient flow evolution that may develop in this situation.

3.3.1. Steady-state flow regime

Consider now the room in Fig. 3.25, which is a simplified generic representation of the International Digital Laboratory mentioned earlier. (Note, the actual building has a number of

storeys each containing its own heat sources, but we take a
simplified view and treat the building as a single space with one
heat source at the base in order to obtain the basic principles
first; we deal with the more complex situation of a multi-storey
space with independent heat sources at different levels later in
Section 5.2.1.) The room in Fig. 3.25 has openings at two levels
connected to the ambient environment and is equipped with
a heated floor. The floor produces a convective current that
rises up the space while mixing the air within. This convective current behaves not dissimilarly to an infinite number of
localised plumes that are positioned closely across the floor
and that quickly merge after rising from their sources (see
the discussion at the end of Section 3.2.5). Depending on the
initial conditions, the convective heating process of this room
varies. We discuss this variation in Section 3.3.2, but for now
we note that the room becomes well-mixed with a well-defined
mean temperature above that of the exterior air at steady
state. At this state, colder ambient air is drawn into the space
through the lower opening while warmer air in the space is
vented out through the upper opening. This steady-state flow
picture, it may be immediately appreciated, is fundamentally
different from that that we have seen associated with the room
heated by a localised source discussed in Section 3.2.3: in the
latter case, the room is stratified into two layers instead of
being well-mixed. Thanks to this well-mixed flow structure,
reduced gravity acts over the entire height of the space in
the case of distributed heating, rather than just over an upper
portion of it as in the case of localised heating. Therefore, if
we denote reduced gravity by g' and the height of the room

by H, we obtain from Bernoulli's theorem an expression for the net volume flux at steady state, Q, in the form

$$Q = A^*(g\alpha\Delta TH)^{1/2} = A^*[g\alpha\,(T_{SS} - T_E)\,H]^{1/2}, \quad (3.3.1\text{-i})$$

where A^* is the effective opening area given by Eq. (2.4-ii) and α is the volume expansion coefficient of air. At steady state, the temperature excess in the room above the exterior air, $(T_{SS} - T_E)$, may be determined from the balance between heat flux produced by the floor and heat loss driven by ventilation. Denoting the former by H_F and the latter by H_V, we obtain from the conservation of thermal energy (cf. Eq. 2.1.2-i)

$$H_F = H_V = \rho C_P Q(T_{SS} - T_E), \quad\quad (3.3.1\text{-ii})$$

where ρ and C_P are the density and specific heat capacity of air, respectively.

Clearly, to estimate the flow rate and interior temperature, the heat flux released by the floor, H_F, must first be identified. This can be done either by looking up in the specifications of the building or, in the case where the building is old and its documentation lost, by taking the temperature of the floor on-site and estimating the heat flux it gives out theoretically. In doing the latter, we may use results from previous studies on turbulent convection from horizontal heated plates, such as those by Townsend (1959), Garon and Goldstein (1973) and Denton and Wood (1979). According to these studies, when the Rayleigh numbers associated with the thermal convection of the floor are of order 10^7–10^9, the heat flux produced by the floor may be related to its surface temperature by the relation

$$H_F = \sigma\,A_F \left(\frac{g\alpha}{\kappa\nu}\right)^{1/3} \rho C_P\kappa(T_F - T_{SS})^{4/3}, \quad (3.3.1\text{-iii})$$

where T_F is the surface temperature of the floor, A_F is the area of the floor, κ is the thermal diffusivity of air and ν is the kinematic viscosity of air. The dimensionless parameter $\sigma = 2^{4/3} C_F$ characterises the heating from the floor, with C_F being the heat transfer coefficient, typically of the value $C_F = 0.066$. For larger Rayleigh numbers, $Ra > 10^9$, Castaing et al. (1989) found that the heat flux H_F scales on $(T_F - T_{SS})^{9/7}$ instead of on $(T_F - T_{SS})^{4/3}$, although the study by Fujii and Imura (1972) suggested that Eq. (3.3.1-iii) is also valid for the range $2 \times 10^8 < Ra < 10^{11}$. The choice of the exponent is unlikely to enter sensitively into calculations, however, because their values are very close $((4/3)/(9/7) = 1.04)$, and any of the aforementioned expressions may be used in practice.

Equations (3.3.1-i) and (3.3.1-ii) enable us to compare the performance of a room heated by a floor heating system with that of a room heated by a localised source such as a small radiator. This comparison is useful because the two forms of heating are probably the most popular in modern buildings. In the case of a localised radiator, combining Eqs. (3.2.3-i) and (3.2.3-ii) given earlier leads to the following expressions for the flow rate and interior temperature:

$$Q = A^{*2/3} \left(\frac{g\alpha}{\rho C_P} \right)^{1/3} H_P^{1/3}(H - h)^{1/3} \quad (3.3.1\text{-iv})$$

and

$$\Delta T = \frac{H_P^{2/3}}{(\rho C_P A^*)^{2/3}(g\alpha)^{1/3}(H - h)^{1/3}}, \quad (3.3.1\text{-v})$$

where H_P is the flux of heating from the radiator, and h is the height of the interface between the upper and lower layers.

As for a heated floor, combining Eqs. (3.3.1-i) and (3.3.1-ii) gives

$$Q = A^{*2/3} \left(\frac{g\alpha}{\rho C_P} \right)^{1/3} H_F^{1/3} H^{1/3} \qquad (3.3.1\text{-vi})$$

and

$$\Delta T = \frac{H_F^{2/3}}{(\rho C_P A^*)^{2/3} (g\alpha)^{1/3} H^{1/3}}. \qquad (3.3.1\text{-vii})$$

Thus, for an identical room geometry and heat input, $H_P = H_F$, the temperature excess in the heated zone due to a localised radiator compared with that due to a heated floor is

$$\frac{\Delta T_{\text{localised radiator}}}{\Delta T_{\text{heated floor}}} = \left(\frac{H}{H-h} \right)^{1/3}. \qquad (3.3.1\text{-viii})$$

And the ventilation volume flux driven by a localised radiator compared with that driven by a heated floor is

$$\frac{Q_{\text{localised radiator}}}{Q_{\text{heated floor}}} = \left(\frac{H-h}{H} \right)^{1/3}. \qquad (3.3.1\text{-ix})$$

Clearly, a radiator produces a warmer heated layer than a heated floor (Eq. (3.3.1-viii)). However, with a radiator, the heated layer occupies an upper portion of the space only, whereas with a heated floor the warm zone extends throughout the height of the space. Thanks to this greater buoyancy head, the ventilation in the room with a heated floor is more rapid (Eq. (3.3.1-ix)). These results have important implications for the selection of the heating system. On the face of it, a localised radiator may seem preferable to a heated floor, owing to the higher temperature it affords in the heated zone and the reduced heat loss it causes through ventilation. However, one problem often experienced in a room heated locally is the feeling of cold feet, which occurs as a result of the settlement of cold ambient air in the lower part of the space, so that the lower leg zone feels chilly even though the upper body zone is warm. We mentioned at the end of Section 3.2.3 that this problem could be addressed by closing more doors and windows and generally reducing means of uncontrolled infiltration, in order to pull down the heated layer to the lower zone. However, if the space were densely occupied and required a relatively large amount of fresh air, this technique could lead to insufficient ventilation. In this situation, an alternative solution could be used; that is, to heat the room with a floor heating system instead. This would lead to the room becoming well-mixed, allowing a comfortable temperature to be achieved throughout the height of the occupied zone, while maintaining good ventilation. The convective heat loss associated with the heated floor could be regulated, and thus energy efficiency optimised, by controlling the vents so as to avoid overventilation while allowing enough fresh air to maintain good indoor air quality.

3.3.2. Evolution to steady state

Let us now look at how the steady state flow picture described in Section 3.3.1 comes about. As mentioned in passing earlier, the convective heating process of a room equipped with a heated floor varies depending on the initial conditions. The study by Fitzgerald and Woods (2007) gives more specificity to this statement: it showed that the form of evolution of the interior environment depends primarily on the relative values of the initial temperature of the room, the temperature of the exterior air and the temperature of the room at the final equilibrium.

3.3.2.1. A room initially at the exterior air temperature or slightly warmer than it

Assume that the exterior temperature is always lower than the interior temperature at the final equilibrium, and that the system is set up so that a comfort temperature is always achieved at the final equilibrium. If the room is initially at the exterior air temperature or slightly warmer than it, distributed thermal plumes generated by the heated floor mix the interior air to a uniform temperature above that of the exterior air soon after the heating has started. At an early stage, the room is cooler than the desired temperature, but as the heating continues, the temperature in the room progressively increases towards the predefined value while the room remains well-mixed (Fig. 3.26). Owing to the well-mixed temperature structure and the progressive heating of the space, the occupants benefit from the heating practically immediately after the heating has started and feel more comfortable as time goes on. The removal of airborne contaminants, too, becomes more effective as the ventilation rate increases as a result of the rise in the temperature of the space.

If we ignore for the moment heat transfer through the room envelope by radiation and conduction (the simplifications adopted throughout this book; see Sections 1.4 and 1.5),

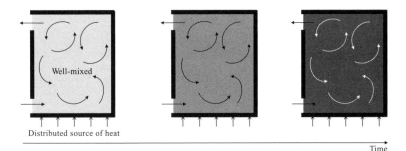

Distributed source of heat

Time

Figure 3.26. Transient evolution of a room with openings at two levels heated by a distributed source at the base, when the room is initially at the exterior air temperature or warmer, but colder than the final equilibrium temperature. (Following Fitzgerald & Woods 2007.)

we may describe the evolution of the interior temperature in terms of the balance between the heating from the floor and heat loss driven by the ventilation; that is,

$$\rho C_P V \frac{d\left(T_{(t)} - T_E\right)}{dt} = H_F - H_V = H_F - \rho C_P A^*$$
$$\times \left(g\alpha H\right)^{1/2} \left(T_{(t)} - T_E\right)^{3/2}, \qquad (3.3.2\text{-i})$$

where V is the volume of the room, $T_{(t)}$ is the interior temperature at time t, T_E is the exterior air temperature, H_F is the heat flux from the floor and H_V is the flux of heat loss driven by the ventilation. The expression (3.3.2-i) indicates a non-linear relation between the rate of heating and the initial temperature in the room. This is shown graphically in Fig. 3.27. It can be seen that a room with a higher initial temperature converges to steady state more quickly. Furthermore, the expression (3.3.2-i) indicates that the time taken to reach steady state increases with increasing room volume. This reflects the fact that it takes longer to heat a greater amount of air contained in

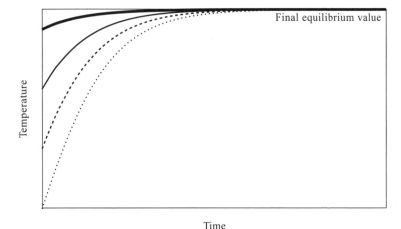

Figure 3.27. Time evolution towards steady state of the temperature in a room heated by a distributed source at the base. The different curves describe the evolution resulting from different initial interior temperatures. (Based on the models by Fitzgerald & Woods 2007 and Gladstone & Woods 2001.)

a larger space. (Note, although room volume affects the evolution time to steady state, it does not affect the temperature at steady state; Eq. (3.3.1-vii).)

3.3.2.2. A room initially warmer than the desired temperature

The evolution of the system will be less straightforward, however, if the room is initially warmer than the desired temperature (Fig. 3.28). The general picture of the transient flow process developed in this situation is as follows. Initial buoyancy associated with warm original air in the room causes an upward displacement flow. The original air is thus drained

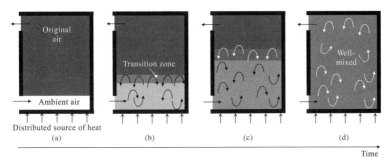

Figure 3.28. Transient evolution of a room with openings at two levels heated at the base by a distributed source, when the room is initially warmer than the final equilibrium temperature. (Following Fitzgerald & Woods 2007.)

upwards and out through the top opening while ambient air is drawn into the room through the bottom opening. On entering the room, the ambient air, being heavier that the existing air in the room, forms a cold layer at the base of the space. This causes an interface to become established between it and a layer of warm original air above (Fig. 3.28a). Initially, the lower layer is cold because the rate of ventilation driven by the hot upper layer is greater than the ventilation rate at the final equilibrium, therefore making the heating from the floor insufficient to raise the temperature of the lower layer to the desired value. However, as the ventilation continues, the temperature of the lower layer increases progressively (Figs. 3.28b, c). Meanwhile, the interface ascends and the upper layer depletes. The ventilation is therefore driven by the buoyancy of the evolving lower and upper layers combined. At the interface, there is penetrative convection caused by the convective current driven by the heated floor. This penetrative convection entrains some air from the upper layer down to the lower layer and creates a transition zone of mixed air of intermediate temperature at the interface (Fig. 3.28b). The general form of penetrative convection developed in this case is comparable to what we have seen driven by a localised source in Sections 3.2.4 and 3.2.5. The obvious difference between the present case and the case involving a localised source is that, with a heated floor, penetrative convection occurs across the interface rather than locally. The volume of the upper layer air entrained by this areal penetrative convection depends chiefly on the temperature contrast between the upper and lower layers and the buoyancy flux of the floor driving the interfacial convection. This entrained volume flux may be quantified using various approaches (see, e.g., Deardorff, Willis & Lilly 1969; Zilintikevich 1991; Lister 1995), but a simple way is to treat the entrainment as bringing to the lower layer a buoyancy flux that is a fraction of the buoyancy flux generated by the floor. It is these buoyancy and volume fluxes entrained by the penetrative convection combined with the buoyancy flux produced by the floor and the building-scale volume flow that govern the rate of ascent of the interface and the evolution of the temperature in the lower zone.

A question of practical relevance may be raised as to how important the mixing zone is to the overall ventilation. An answer to this is that it depends on the depth of the mixing zone relative to the height of the space and the vertical position of the occupied zone. Fitzgerald and Woods (2007) calculated that a warm space of height 10 m heated by a uniform load of order 30 W/m^2 and ventilated at a typical rate of about three to four air changes per hour will have a mixing zone about 1 m deep developing over a time of about 15–30 minutes, the system's ventilation timescale. Penetrative convection that creates the mixing zone in this case is not insignificant but in a building of vertical extent 10 m or more is dominated by the effects of floor heating and the upward displacement flow. Thus the space effectively becomes stratified into two distinct layers. On the contrary, in spaces of smaller vertical extent, penetrative convection may have a greater impact on the overall flow, causing the room to become more smoothly stratified and the ventilation rate and comfort perception to be modified accordingly.

Fig. 3.29 shows how the initial temperature in the room affects the evolution of the interior conditions and comfort sensation. For a room initially *much hotter* than the final steady-state temperature, the ventilation driven by the upper layer dominates the heating in the lower layer. Consequently, the flow is initially rapid and the lower layer deepens quickly (thick solid line in Fig. 3.29). Due to the rapid ventilation, the lower layer attains a relatively low temperature initially (thick dashed line in Fig. 3.29). Extreme discomfort may thus be experienced, depending on the exterior temperature. However, as the upper layer drains, the overall buoyancy driving the ventilation reduces. This causes the rate of ascent of the interface to decrease (thick solid line in Fig. 3.29) and the lower layer to heat up increasingly rapidly (thick dashed line in Fig. 3.29). Comfort is therefore approached with increasing speed. Thanks to the strong initial temperature contrast between the upper and lower layers, penetrative convection

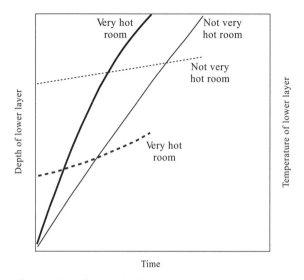

Figure 3.29. Time evolution of the depth of the lower layer (solid lines) and its temperature (dashed lines) in a room initially warmer than the final equilibrium temperature. (Based on calculations by Fitzgerald & Woods 2007.)

in this case is negligible and the zone of mixed air at the interface is thin, hardly affecting the comfort level in the lower zone.

In contrast, if the room is initially only *slightly warmer* than the final equilibrium temperature, the ventilation rate is small, so that the heating by the floor becomes significant. Consequently, the interface ascends more slowly (thin solid line in Fig. 3.29), and the lower layer attains a higher temperature initially (thin dashed line in Fig. 3.29). The occupants therefore feel more comfortable at the beginning. In this situation, because the temperature contrast between the two layers is small, penetrative convection is significant. As a result, there is a deep region of mixed air at the interface. If this mixing region lies within the occupied zone, it will allow the occupants to benefit more from warm air convected from the upper layer. In extreme cases, an extended mixing zone may develop resulting in a well-mixed interior. This is even more conducive to thermal comfort.

It should be noted that, in any of the above cases, the period of discomfort could extend almost throughout the occupancy session if the windows were left constantly wide open. This is because the resultant fast ventilation flow would prevent the incoming ambient air from heating up significantly. To overcome this problem and to keep the temperature in the lower occupied zone as close to the comfort temperature as possible, the openings should be gradually opened to allow sufficient time for the air in the lower layer to heat up. This operation could be accomplished through a computerised building management system, which links the actuators of the openings with readings of the temperature in the lower zone.

Regardless of the severity of the initial interior temperature, once the interface rises to the top of the space, the original air in the room is completely drained and so the room becomes well-mixed (Fig. 3.28d). From this point on, the entire room heats up progressively in the same manner as that of the room starting off at the same temperature as the exterior air or slightly warmer than it. This heating process has been described in Section 3.3.2.1.

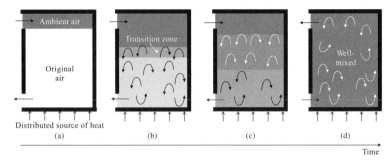

Figure 3.30. Transient evolution of a room with openings at two levels heated at the base by a distributed source, when the room is initially colder than the exterior. (Following Fitzgerald & Woods 2007.)

3.3.2.3. A room initially colder than the exterior air temperature

The picture of the transient evolution process will be different from that described in Section 3.3.2.2 if the room is initially colder than the exterior air temperature. Now negative buoyancy associated with cold air in the room drives the flow downwards initially. The original air in the room thus drains through the bottom opening while the exterior air enters through the top opening. At an early time, the inflowing exterior air is lighter than the original air in the room, and so it ponds above, causing the room to become stratified into two layers (Fig. 3.30a). Owing to this stratification, the occupants in the lower zone may feel uncomfortably cold initially. However, as the ventilation proceeds, the temperature in the lower zone increases towards the comfort value, thanks to the heating from the floor. Meanwhile, the interface between the two layers descends and the lower layer depletes (Figs. 3.30b, c). This depletion of the lower layer combined with the heating from the floor causes the rate of downward draining to decrease with time. At the interface, thermal plumes produced by the heated floor drive penetrative convection in a manner similar to that observed in the room of Section 3.3.2.2. This penetrative convection, adding heat to the lower zone and producing an intermediate layer of positively buoyant warm air, also contributes to the slowing down of the descent of the interface. As the temperature in the lower layer rises closer to that of the upper layer, the penetrative convection becomes more significant, leading to the deepening of the intermediate zone (Figs. 3.30b, c).

Depending on the initial temperature of the room, the penetrative convection may be sufficiently significant that it greatly affects the overall flow evolution and the perception of indoor thermal comfort. Figure 3.31 explains this in more detail. It shows that if the room is initially *very cold*, the draining is rapid and the lower layer is almost completely drained before it is heated to near the temperature of the upper layer (thick solid line and thick dashed line in Fig. 3.31). In this situation, the intermediate mixing zone is thin, and penetrative convection becomes significant only just before the point at which the interface reaches the floor. At this point, there is a rapid overturn whereby the whole room becomes well-mixed and heats up quickly to the exterior air temperature (see the end of the thick dashed line in Fig. 3.31). The heating of the room after this point follows the process described

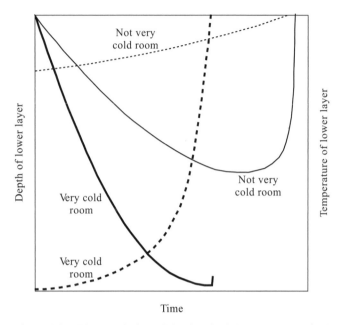

Figure 3.31. Time evolution of the depth of the lower layer (solid lines) and its temperature (dashed lines) in a room initially colder than the exterior. (Based on calculations by Fitzgerald & Woods 2007.)

in Section 3.3.2.1. Thanks to the initial rapid ventilation and the fast heating of the space after the overturn, the occupants feel comfortable relatively quickly after the ventilation has started. Moreover, old air in the room is replaced quickly, helping to prevent an excessive accumulation of heat/contaminants.

On the contrary, if the room is initially *not very cold*, negative buoyancy is weak and the draining is slow. This allows more time for the lower layer to be heated towards the temperature of the upper layer (thin dashed line in Fig. 3.31). The small temperature difference between the upper and lower layers, in turn, leads to significant penetrative convection and a deep transition zone at the interface. As the ventilation continues, the lower layer heats up further. This causes the transition zone to deepen, allowing the plumes from the floor to rise higher in the space (Fig. 3.30c). Meanwhile, the lower layer depletes, but at a slower rate than in the room starting off at a very low temperature. This depletion of the lower layer continues until the interface reaches a minimum height, after which the lower layer begins to deepen (thin solid line in Fig. 3.31). By this point, the mixing zone is already extensive. Eventually, when the temperature of the lower layer increases to that of the exterior environment (and the upper layer) the room becomes well-mixed, and an upward displacement flow develops. The temperature in the room then increases progressively towards the required value, following the process described in Section 3.3.2.1. Thanks to the deeper mixing zone, the occupants in the room subjected to this transient flow regime experience a warmer environment than those in the room with a very low initial temperature. However, with the room not very cold originally, the residual air in the lower occupied zone may stay at low levels for a considerable period of time before the overturn kicks in. This delays the approach of thermal comfort and may cause a significant build-up of CO_2 levels, leading to poor indoor air quality. To provide sufficient fresh air and speed up the attainment of comfort conditions, the space may need to be over-ventilated initially to accelerate the draining and the commencement of the upward displacement flow. This may be achieved by regulating the openings appropriately.

3.4. A combination of the localised source and the distributed source

Many modern buildings are subjected to a combination of localised and distributed heating/cooling. A heated floor may be used in a space occupied locally, or workers may be distributed across an office floor while printers and photocopiers are clustered in one spot. In a theatre, the audience spreading out in the main body of the space may act as a distributed source of heat while a concentrated group of actors and lighting on the stage may act as a localised source of heat (Fig. 3.32). The basics obtained from previous sections of the behaviours of localised and distributed sources of buoyancy make possible an examination of the mechanics of flows developed in these situations. Consider the room in Fig. 3.33. This room has openings at two levels and is heated by a localised source and a distributed source at the base. The system was first studied by Hunt, Holford and Linden (2001), who carried out water-bath experiments using a combination of a hot water plume and a hot plate. The problem was later revisited by Chen-vidyakarn and Woods (2008) in the context of underfloor air-conditioning/mechanical air supply; a pump was used to drive flows in their experiments instead of pure buoyancy, equivalent to using a fan to drive air in a full-scale building. The flow principles acquired from this latter study, however, carry through to the case of natural ventilation; a change in the ventilation rate effected by varying the speed of a fan in a mechanically ventilated space may be accomplished through an adjustment of a window in a naturally ventilated space. It

Figure 3.32. An example of a space containing combined localised and distributed heating/cooling. In a theatre, actors and lighting on the stage may together be regarded as a localised source of heat while the audience distributed in the main body of the building may be regarded as a distributed source of heat.

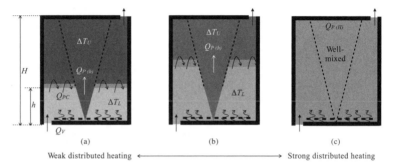

Figure 3.33. Impact of the relative strength of distributed and localised heating on the flow structure. (Following Chenvidyakarn & Woods
2008.)

is the findings from the above two studies that will form the
basis for much of our discussion below.

To appreciate what happens in a room with combined
localised and distributed heating/cooling, it is useful to recall
first the classic ventilation of a plume described by Linden,
Lane-Serff and Smeed (1990; see Section 3.2.3). In this problem, a room with vents at two levels heated by a single localised source at the base becomes stratified into two layers
at steady state, with the upper layer supplied by the localised plume attaining a temperature greater than that of the
exterior air, and the lower layer supplied by fresh ambient air
maintained at the exterior temperature. A key result of this
work is that the height of the interface between the two layers
is independent of the magnitude of heating but controlled by
the geometry of the space (Eq. (3.2.3-iii)).

However, work by Hunt, Holford and Linden (2001) and
Chenvidyakarn and Woods (2008) showed that if the room
also contains a distributed source of heat at the base, the

relative magnitude of localised and distributed heating enters into the determination of the interface height. If the magnitude of distributed heating is sufficiently large compared with that of localised heating, the room becomes well-mixed at steady state, akin to the room heated solely by a distributed source discussed in Section 3.3.1 (cf. Fig. 3.25 and Fig. 3.33c). A key departure of the present case from the case of a sole distributed source is that, with a combination of a localised and a distributed source, the heat fluxes from both sources contribute to the temperature of the room. Therefore, if we let H_D represent the flux of distributed heating and H_P the flux of localised heating, we may write the temperature excess in the well-mixed room above the exterior in the form

$$\Delta T_{\text{well-mixed}} = \frac{H_D + H_P}{\rho C_P Q_V},\qquad(3.4\text{-i})$$

where ρ and C_P are the density and specific heat capacity of air, respectively. The net volume flux Q_V is driven by the reduced gravity acting over the entire height of the room and, by virtue of the conservation of mass, must equal the volume flux of the plume at the level of the upper opening, $Q_{P(H)}$. Thus we may write for the well-mixed room

$$Q_{V,\text{well-mixed}} = Q_{P(H)} = \lambda \left(\frac{g\alpha H_P}{\rho C_P} \right)^{1/3} H^{5/3}$$
$$= A^*(g\alpha \Delta T_{\text{well-mixed}} H)^{1/2},\qquad(3.4\text{-ii})$$

where $Q_{P(H)}$ is given by the plume theory (Eq. (3.2.1-xiii)), with λ an integrated constant describing the entrainment of the plume. The variable H is the vertical separation between the upper and lower openings, A^* is the effective opening area,

g is gravitational acceleration and α is the volume expansion coefficient of air.

However, if localised heating is sufficiently strong compared with distributed heating, the flow structure is quite different. Now, instead of being well-mixed, the room is stratified into two layers, not dissimilar to the room heated by a localised source alone. The key distinction between the present room and the room heated solely by a localised source is that, with a distributed source also present in the space, the lower layer is heated uniformly, and attains a temperature greater than that of the ambient environment rather than one identical to it (Fig. 3.33a). Therefore, the ventilation flow is driven by the reduced gravity acting over the two buoyant layers combined. If we let ΔT_U be the temperature excess in the upper layer above the exterior air, ΔT_L the temperature excess in the lower layer above the exterior air, and h the height of the interface between the two buoyant layers, we may write for the ventilation volume flow of the stratified room

$$Q_V = A^*[g\alpha(\Delta T_L h + \Delta T_U (H - h))]^{1/2}. \quad (3.4\text{-iii})$$

At the interface between the two layers, there is areal penetrative convection similar to that observed in Sections 3.3.2.2 and 3, which is driven by the heating from the floor. This penetrative convection produces a transition zone of mixed air at the interface. This transition zone, however, may be treated implicitly as part of the upper layer if it is sufficiently thin. This simplification allows Eq. (3.4-iii) to hold and calculations to be simplified. To quantify the volume flux of entrainment associated with the penetrative convection the approach due to Zilintikevich (1991) may be used. This treats the entrainment as convecting from the upper layer to the lower layer a fraction k of the buoyancy flux produced by the floor which drives the convection at the interface. Thus the volume flux of entrainment, Q_{PC}, may be given by

$$Q_{PC} = \frac{kH_D}{\rho C_P(\Delta T_U - \Delta T_L)}. \qquad (3.4\text{-iv})$$

This volume flux of entrainment modifies the heat and mass balances of the system, and so for the upper layer the net volume flux Q_V becomes equal to the volume flux $Q_{P(h)}$ supplied by the plume at the level of the interface less the volume flux Q_{PC} taken away by the penetrative convection,

$$Q_V = Q_{P(h)} - Q_{PC}, \qquad (3.4\text{-v})$$

where $Q_{P(h)}$ is given according to Eq. (3.2.1-xiii) as

$$Q_{P(h)} = \lambda \left(\frac{g\alpha H_P}{\rho C_P} \right)^{1/3} h^{5/3}. \qquad (3.4\text{-vi})$$

Furthermore, the conservation of thermal energy requires that, in each layer, the total heat input balances the net rate of advection associated with that layer. The lower layer is heated by a combination of the distributed source and heat received from the upper layer through the penetrative convection. Therefore, its temperature is given by

$$\Delta T_L = \frac{H_D + \rho C_P Q_{PC} \Delta T_U}{\rho C_P Q_{P(h)}}. \qquad (3.4\text{-vii})$$

The upper layer, on the other hand, is heated by the localised plume which carries heat flux directly from the source, and

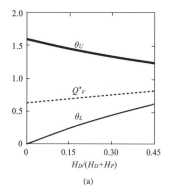
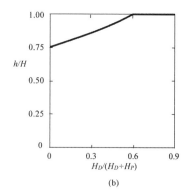

(a)   (b)

Figure 3.34. Impact of the relative magnitude of distributed heating, $H_D/(H_D + H_P)$, on the dimensionless volume flow rate Q^*_V (a; dashed line), the dimensionless temperature in the upper layer, θ_U (a; solid thick line), the dimensionless temperature in the lower layer, θ_L (a; solid thin line) and the dimensionless depth of the lower layer, $h^* = h/H$ (b). (Based on model by Chenvidyakarn & Woods 2008.)

which has entrained warm air from the lower layer. Therefore, its temperature is given by

$$\Delta T_U = \frac{H_D + H_P}{\rho C_P Q_V} = \frac{H_D + H_P}{\rho C_P (Q_{P(h)} - Q_{PC})}. \quad \text{(3.4-viii)}$$

The transition from the stratified room to the well-mixed room deserves some attention, as this describes what essentially happens in the building when its occupancy changes. Analysis of this may be carried out using Eqs. (3.4-i)–(3.4-viii); the results are shown in Fig. 3.34. Note, in this figure, the flow rate, the temperatures and the depth of the lower layer are shown in dimensionless terms as scaled on those of the well-mixed room, as $Q^*_V = Q_V/Q_{P(H)}$, $\theta_U = \Delta T_U/\Delta T_{\text{well-mixed}}$, $\theta_L = \Delta T_L/\Delta T_{\text{well-mixed}}$ and $h^* = h/H$, respectively. Also, a typical value of $k = 0.1$ is used in the calculations for illustration purposes. It can be seen that if the net heat flux $(H_D + H_P)$ is fixed, an increase in the relative strength of distributed heating, $H_D/(H_D + H_P)$, raises the temperature of the lower layer (Fig. 3.34a, thin solid line). This, in turn, increases the overall buoyancy drive, leading to a faster flow (Fig. 3.34a, dotted line). However, because the net heat flux to the upper layer is fixed at $(H_D + H_P)$, the increase in the ventilation flow causes the temperature of the upper layer to decrease (Fig. 3.34a, thick solid line). Meanwhile, the increase in the heating from the floor makes the convective current produced by the floor become more vigorous, causing greater entrainment of air from the upper layer to the lower layer. As a result, the lower layer deepens, $h/H \to 1$ (Fig. 3.34b). Eventually, when the relative heating from the floor is increased sufficiently, the lower layer deepens to the level of the upper opening, $h/H = 1$. At this point, the room becomes well-mixed (Fig. 3.34b).

An examination of the mathematical model given in the preceding text will show that the critical relative magnitude of distributed heating that causes the room to become well-mixed is sensitive to the value of k describing the strength of penetrative convection. A larger k leads to a deeper lower layer and therefore a smaller critical relative flux of distributed heating. This statement is described graphically in Fig. 3.35. The figure shows that for k within a typical range of 0.05–0.2 reported by Zilintikevich (1991), the critical relative flux of distributed heating lies between $0.47 \leq H_{D\,crit}/(H_D+H_P) \leq 0.68$. In other words, a distributed heat load roughly half the total heat load could lead to the room becoming well-mixed.

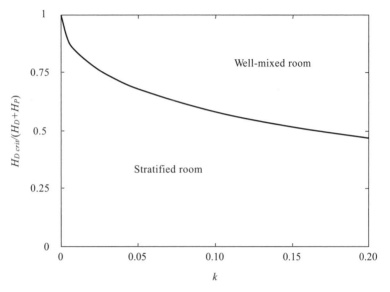

Figure 3.35. Impact of the penetrative convection parameter k on the critical relative magnitude of distributed heating, $H_{D\,crit}/(H_D + H_P)$. (Based on model by Chenvidyakarn & Woods 2008.)

The principles outlined above find a particular application in the control of spaces such as an auditorium (Chenvidyakarn & Woods 2008; Fig. 3.32). As mentioned earlier, the audience in the main body of the building may be treated as a distributed source of heat while the actors and lighting on the stage may be treated as a localised source. A solution may be found of the appropriate effective vent size which is required to maintain the lower occupied zone at a comfortable temperature while keeping any pocket of warm air well above the heads of the occupants when there are certain heat loads in the building. For example, a small auditorium of volume about 3000 m³ and height 10 m may contain 8000 W in

the seating and 7500 W on the stage. Assume that this space
needs to be ventilated at the rate of at least one air change
per hour; that any pocket of warm air needs to be kept at
least 7 m from the floor to be well clear of the balcony audience; and that the temperature in the lower zone needs to be
maintained within 5°C above the exterior air temperature to
prevent overheating. The above model gives the appropriate
effective vent size for this space as about 2 m^2. This value of
the appropriate vent size varies with the geometry of the space
and the magnitudes of the internal heat loads. An effective
vent size of almost 3 m^2 may be required instead if the heat
load in the seating is changed to 10000 W and that on the
stage is changed to 5500 W, for instance. A similar analysis
may be performed on other kinds of space, such as an office,
to determine the effective vent size required to keep a pocket
of warm air produced by a cluster of printers and photocopiers above the working zone, for example. The results would
help inform design at an early stage as well as the control of
the space during operation.

4   Sources of opposite sign

Besides providing sufficient fresh air, another major goal of natural ventilation is to maintain indoor air temperature within comfort limits. For buildings in colder climates and those with smaller heat gains, this goal may be achieved through manipulation of internal sources of heat alone, using perhaps the basic principles discussed in Chapter 3. However, for buildings in warmer climates and those with larger heat gains, incorporation of some form of cooling may be necessary. In naturally ventilated spaces this cooling may be provided mechanically, through a chilled ceiling, for example, or passively, through thermal mass, for example.

Dynamics-wise, a flow produced by a source of cooling is identical to that produced by a source of heating of the same geometry; one is just an invert of the other. The basic principles presented in Chapter 3, therefore, carry through to this chapter. However, the inclusion of cooling does present a challenge for ventilation control: air warmer than the ambient environment rises whereas that colder falls, giving rise to competition between positive buoyancy associated with the warm air and negative buoyancy associated with the cold air. This competition slows down the flow, potentially leading to insufficient ventilation and thermal discomfort. To prevent these problems, careful balancing between positive and negative buoyancy is required. The key to achieving this is insight into the interactions between the two types of buoyancy.

In this chapter we attempt to develop this insight. We begin by examining the case in which air that has been pre-cooled outside the ventilated space, for instance, by being passed through a chiller or a cool thermally massive material, is used to flush a warm, unoccupied space containing no steady source of heat. Then, we consider the case of a steadily occupied space ventilated with pre-cooled air, before turning to flows driven by cooling within the space itself, such as those encountered in a building containing a chilled ceiling or internally exposed thermal mass. It will be seen that incorporation of cooling can lead to unexpected flow phenomena, such as a rise in the interior temperature after an increase in cooling input and flow oscillations during the draining of a space.

4.1. Flushing with pre-cooled air

We have considered the flushing of a space containing residual heat using untempered ambient air in Section 3.1. When this ventilation technique is used, the intention is usually that after a period of flushing the room becomes sufficiently comfortable for occupation, and that the time required for the room to attain thermal comfort is not excessive. However, thermal comfort will not be achieved if the temperature of the exterior air lies above a comfort limit, as may be the case in summer or in warmer climates; it may be recalled that the lowest temperature achievable in a space flushed in this manner is the same as that of the ambient air. In this situation, an alternative technique may be employed that uses air that has been pre-cooled to a sufficiently low temperature, by passing through a cooling medium such as a massive floor slab, a rock bed or a cooling coil, to flush the space instead. This pre-cooled air may be supplied to the space through an undercroft/raised floor space and distributed to the main occupied zone through diffusers on the floor or, in the case of a cinema and other similar spaces, through outlets located underneath the seating. To make the set-up cost-effective the undercroft/raised floor space may be made quite deep to serve a dual role as a storage or mechanical space. To conserve energy, naturally created thermal buoyancy (alone or in combination with wind) may be used in place of an electric fan to drive ambient air through the cooling medium, and to push the resultant cooled air through the occupied zone. This technique is in fact not new. It dates back to ancient Persia and Rome (Fig. 4.1), and occasionally reappears in modern buildings, such as at the Braunstone Health & Social Care Centre in Leicester (Fig. 4.2) and the Lanchester Library in Coventry, both in the UK (Fig. 4.3; see Cook, Lomas & Eppel 1999). Further, the technique may be used not only to flush spaces containing residual heat/contaminants, but also to provide cooling in spaces steadily occupied. In the latter case, there is a great opportunity for energy saving, especially when the technique is employed in cinemas, offices and other similar buildings

Figure 4.1. Left: A typical traditional Persian building, as seen from the outside, showing its wind towers. Right: Its naturally operated underground cooling system. External air is drawn into the building through underground channels that cool the air as it passes. The flow is driven by a combination of thermal buoyancy and wind. The latter is manipulated through the wind towers.

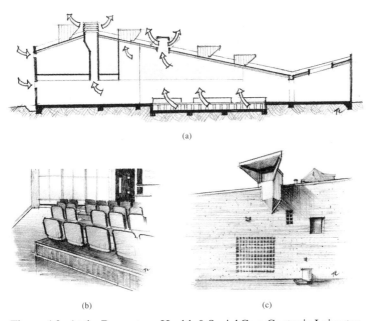

(a)

(b)   (c)

Figure 4.2. At the Braunstone Health & Social Care Centre in Leicester, UK, fresh air is supplied pre-cooled through an underground plenum (a) based on the idea of the Roman hypocaust. The cooled air is then distributed to the interior space through openings located underneath the seating in the waiting hall (b). Warm foul air is vented from the space through high-level stacks (c).

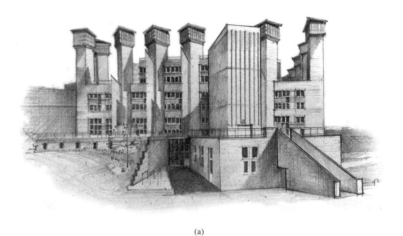

(a)

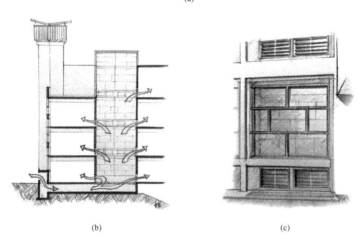

(b)   (c)

Figure 4.3. The Lanchester Library in Coventry, UK (a) uses underground plenums (b) to supply pre-cooled air to the reading area via openings located on the walls of its atria (b, c).

that usually use central air conditioning. This is because typically about 70–80% of the energy delivered to central air conditioning is spent in operating its fans (Neufert 1980), and this portion of energy can be saved by using naturally created

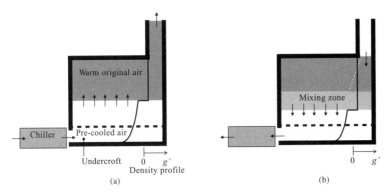

Figure 4.4. The flushing of a warm space with pre-cooled air during a period of upward draining (a) and a period of downward draining (b). (Following Chenvidyakarn & Woods 2006.)

thermal buoyancy to drive air instead. We discuss the cooling of steadily occupied spaces in Section 4.2, but for now let us focus on the flushing of warm but empty spaces.

The process and mechanism of 'pre-cooled flushing', as the technique may be called, were first explained by Chenvidyakarn and Woods (2006). The essence of their findings is shown by the simple diagrams in Fig. 4.4 and the series of shadowgraph images in Fig. 4.5. Suppose a space is equipped with an undercroft that is connected to a 'chiller'. This chiller, taking the form of an underground thermal mass labyrinth, a rock bed, a cooling coil or some other similar cooling device, allows fresh air to pass and be chilled before being supplied to the main space via the undercroft. Let us suppose also that the space has a ventilation chimney or 'stack' at the top, which opens directly onto the exterior environment and therefore facilitates ventilation. (We clarify at the end of this section why this set-up is chosen instead of the usual flush opening

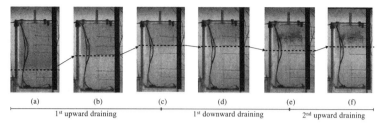

(a)   (b)   (c)   (d)   (e)   (f)

1st upward draining   1st downward draining   2nd upward draining

Time

Figure 4.5. Shadowgraphs showing the evolution of the interface height during the initial upward draining (a–c), the first downward draining (c–e) and the second upward draining (e–f), as observed in an experiment by Chenvidyakarn and Woods (2006). The inflowing fresh fluid from the stack is dyed dark. (Adapted from Chenvidyakarn & Woods 2006.)

used in our other analyses, but for now let us concentrate on the flow process.) Imagine, then, that after a period of occupancy the space is left uncomfortably warm (and perhaps full of contaminants) and needs to be flushed before it can be reoccupied. Positive buoyancy produced by warm residual air in the space drives flow upwards, so that warm air drains through the stack while fresh air is drawn into the space, first through the chiller, then through the undercroft (Fig. 4.4a). As in the case of flushing with no cooling discussed in Section 3.1.2, fresh air, being heavier than residual air, establishes a layer in the lower part of the space. This layer is separated from a layer of residual air above by an interface. This interface ascends as the room drains upwards (Figs. 4.5a, b). However, in contrast to the space flushed with untempered air, the space equipped with a pre-cooling system receives fresh air that has been chilled to a temperature below that of the ambient environment. Therefore, the air in the lower layer has negative buoyancy associated with it. This negative buoyancy competes with the positive buoyancy of the original air in the upper layer. Initially, the lower layer is shallow and the upper layer deep, and so the draining is rapid. However, as the upper layer depletes and the lower layer deepens, the positive buoyancy head decreases while the negative buoyancy head increases. This causes the flow to slow down and the interface to ascend at reduced speed. This slowing of the flow allows more time for the intake air to sit in the chiller and become cooler with time. Consequently, the lower layer becomes stably stratified according to the density profile sketched in Fig. 4.4a. This stratification strengthens negative buoyancy in the lower layer. Eventually, the upward draining stops when the interface rises to a height such that the positive buoyancy of the upper layer becomes matched by the negative buoyancy of the lower layer. At this point, the interface is arrested at a height above the base of the room, even though the original air has not been completely drained (Fig. 4.5c).

At equilibrium, the air in the upper part of the room is still warmer than the ambient air, and so the interface between these two fluids at the level of the stack opening is unstable (i.e., it is subjected to Rayleigh–Taylor instability; see Section 2.2). This soon causes the ambient air to migrate into the stack, which in turn leads to a decrease in temperature in the stack. This upsets the balance between the positive and negative buoyancy held at the point of equilibrium. Consequently, a period of downward draining ensues (Figs. 4.4b and 4.5c–e). During this period, the ambient air enters the room through the stack, mixes with the air in the upper layer and descends to form an intermediate layer above the cold lower zone. This intermediate layer deepens as the draining progresses because of entrainment of air in the upper layer. However, if mixing between the inflowing ambient air and the air in the upper layer is vigorous, the intermediate layer may be treated as part of a well-mixed upper layer in simple analysis. Initially, the downward draining is fast because the lower layer is relatively deep and strongly stratified. However, as the room drains, the lower layer depletes and the upper layer deepens. This causes a decrease in negative buoyancy and an increase in positive buoyancy. As a result, the downward draining slows down. Eventually, the flow ceases when the lower layer depletes to a height such that the negative buoyancy associated with it becomes balanced again by the positive buoyancy associated with the upper layer (Fig. 4.5e). At this point, the interface is arrested closer to the base of the room than at the previous equilibrium.

As soon as the downward draining stops, warm air in the upper layer, being lighter than the ambient air in the stack above, flows back up the stack. This causes an increase in the head of positive buoyancy and, consequently, the space begins to drain upwards again (Fig. 4.5f). The upward draining process described earlier is then repeated, albeit with a slower initial flow, owing to a lower initial temperature in the upper layer caused by the mixing between the ambient air and the warm air in the upper layer during the preceding downward draining session. After a period of upward draining the system converges to equilibrium again once the positive buoyancy in the upper part of the room and the negative buoyancy in the lower part of the room become matched. At this point, the interface is arrested above its previous position. Rayleigh–Taylor instability at the stack opening then causes downward draining to resume, and the downward draining process described earlier is repeated. In this way, the pre-cooling causes the system to oscillate between upward and downward draining cycles, while the temperature profile of the air in the room evolves progressively towards that of the ambient environment (i.e., it becomes less stratified and increasingly neutrally buoyant). Finally, the ventilation ceases altogether once the room is completely drained and attains the temperature of the exterior air.

The described oscillatory draining process has direct implications for occupancy comfort. To appreciate this, consider the sketch in Fig. 4.6. The figure shows schematically the evolution of the temperature at a height within the lower zone, which is likely to be occupied, just below the level of the interface at the first equilibrium. This temperature (traced by the solid line in the figure) fluctuates about the temperature of the ambient air, becoming lower when a period of upward draining causes the interface to ascend past the position in question and higher when a period of downward draining causes the interface to descend below it. In contrast to flushing without pre-cooling (represented in the figure by the dashed line), the lowest temperature achievable in the room equipped with pre-cooling is below that of the ambient air rather than equal to it. This minimum temperature is reached at the first equilibrium when the height of the interface is also the greatest. For this reason, the conditions at the first equilibrium often

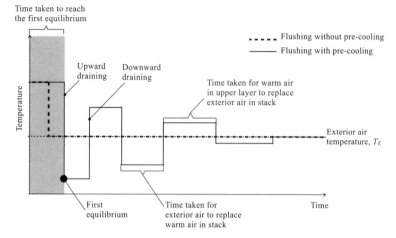

Figure 4.6. Conceptual sketch of the evolution of the interior temperature at a height within the lower occupied zone just below the position of the interface at the first equilibrium. The solid line represents a room flushed with pre-cooled air. The dashed line represents a room flushed with untempered ambient air. (Adapted from Chenvidyakarn & Woods 2006.)

determine the effectiveness of flushing, while the draining time to converge to the first equilibrium often influences the occupancy pattern.

A key question is how to control the system to achieve a desired temperature in the lower occupied zone within an acceptable span of time. It has been shown by Chenvidyakarn and Woods (2006) that the conditions at the first equilibrium are governed by the balance between the reduced gravity associated with cold air from the chiller and the reduced gravity associated with warm residual air. This balance may be described in terms of parameter

$$G = \frac{g'_L(0,0)}{g'_U}, \qquad (4.1\text{-i})$$

where g'_U is the initial buoyancy of the upper layer, and $g'_L(0,0)$ is the initial buoyancy of the cold lower layer at the point of inflow (i.e., at time $t = 0$ and height $z = 0$). The initial buoyancy $g'_L(0,0)$ is related to the initial cooling flux $H_C(0)$ of the chiller and the initial flow rate $Q(0)$ by the relation

$$g'_L(0,0) = \frac{g\alpha H_C(0)}{\rho C_P Q(0)}, \qquad (4.1\text{-ii})$$

where g is gravitational acceleration, α the volume expansion coefficient of air, ρ the density of air and C_P the specific heat capacity of air. The initial flow rate $Q(0)$ is estimated at the point before the flushing starts, that is, when the room is well mixed and full of hot air. This can be done using Eq. (3.3.1-i), with ΔT defined as $(T_{\text{residual}} - T_E)$. The initial cooling flux is given in terms of the initial volume flow rate, the initial temperature of the chiller, the properties of air and the geometry of the chiller as

$$H_C(0) = \beta Q(0)^m (T_E - T_C(0)), \qquad (4.1\text{-iii})$$

where T_E is the temperature of the ambient air, $T_C(0)$ is the initial temperature of the chiller, and β is an integrated constant describing the properties of air and the geometry of the chiller, whose value varies from case to case. The value of the exponent m depends on the behaviour of flow through the chiller, which dictates the effectiveness of heat transfer between the chiller and the air that comes into contact with it. Flows driven by buoyancy are usually rather slow but can be either laminar or turbulent, depending greatly on the geometry of the chiller. For example, a chiller whose geometry may be considered a flat plate or a combination thereof (e.g., one made up of concrete slabs) may have a critical Reynolds number (below which the flow is laminar and above which the flow is turbulent) between 5×10^5 and 10^6, whereas a chiller whose geometry resembles that of a tube (e.g., an earth pipe or a hollow core slab) may have a lower critical Reynolds number between 2×10^3 and 4×10^3 (Holman 1997). A buoyancy-driven flow of typical speed less than 1 m/s over a distance of 10 m may be laminar in the former but turbulent in the latter. Churchill and Ozoe (1973) showed that if the flow through the chiller is laminar, H_C scales on $Q^{1/2}$. Holman (1997) showed that if the flow is turbulent, H_C scales instead on $Q^{4/5}$. These relations suggest that the type of chiller plays a significant role in determining the amount of cooling obtained, and that the effect of this is amplified with an increase in the flow rate.

To illustrate the impact of the parameter G on the ventilation flow, a series of sketches are shown in Fig. 4.7, based on the model by Chenvidyakarn and Woods (2006). These sketches show how variation in the value of G affects the draining time to converge to the first equilibrium (a), the height of the interface at the first equilibrium (b) and the temperature structure at the first equilibrium (c). It can be seen that for small G, corresponding to large openings or weak cooling, there is relatively rapid flow (Fig. 4.7a) and so little time for fresh air to be cooled by the chiller. The room is thus relatively weakly stratified (Fig. 4.7c), with the interface arrested nearer to the ceiling (Fig. 4.7b). In addition, the draining time is relatively long (Fig. 4.7a). On the contrary, for large G, corresponding to small openings or strong cooling, there

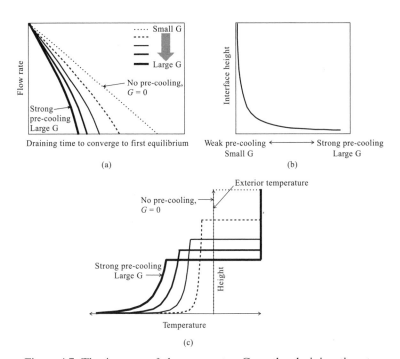

Figure 4.7. The impacts of the parameter G on the draining time to converge to the first equilibrium (a), the interface height at the first equilibrium (b) and the interior temperature profile at the first equilibrium (c). (Based on model by Chenvidyakarn & Woods 2006.)

is relatively slow flow (Fig. 4.7a) and hence a relatively long time for fresh air to be cooled by the chiller. Consequently, the room is relatively strongly stratified (Fig. 4.7c), with the interface arrested closer to the floor (Fig. 4.7b). The draining time in this case is relatively short (Fig. 4.7a).

The outline of flow behaviour in Fig. 4.7, it must be noted, is based on the assumption adopted by Chenvidyakarn and Woods (2006) that the temperature T_C of the chiller remains constant throughout the ventilation process. This is a reasonable assumption for a chiller that is mechanical in nature, such as a cooling coil operated by a compressor, whose temperature may be arbitrarily controlled. However, if the chiller takes the form of thermal mass, its temperature will evolve over time as the mass exchanges heat with the intake air. To be strictly correct, the temperature evolution of the mass must be considered in order to arrive at an accurate value of T_C at a particular time point. This will indeed be a necessary procedure when analysing a system equipped with a smaller thermal mass, whose temperature evolves appreciably over the timescale of draining. However, the procedure may be ignored for larger thermal mass whose temperature deviates so little over the draining period that it may be regarded as constant. In the case of small thermal mass, the broad-brush flow picture will be as sketched earlier, but the lower layer will not be so strongly stratified and the flow not so slow late in the ventilation process, owing to the mass heating up, reducing negative buoyancy. In an extreme case in which the temperature of the mass reaches that of the ambient air halfway through the draining process, the oscillation will cease altogether even though the residual air has not been completely drained. From this point on, the room will drain in a simple upward manner not dissimilar to the draining of a room with no cooling discussed in Section 3.1.2 (with a key difference being that in the present system the draining will be slower as a result of the negative buoyancy associated with the layer of cold air established when the cooling was still in operation).

The general flow principles described in the preceding text provide a basis for effective control of the system.

In practice, the key aim of flushing a space is usually to achieve indoor thermal comfort within the shortest period of time possible to minimise the turn-around time of occupancy. This would likely be important in spaces such as cinemas and theatres in particular, where the number of rounds of performance is directly linked to profit. In these cases, the challenges lie primarily in determining the appropriate amount of cooling, the appropriate effective vent size and the time taken to flush the space. For a given space and a given chiller we may use a plot similar to Fig. 4.7c to explore variation in the interior temperature structure at the first equilibrium in relation to the value of the parameter G. A complete mathematical model for this undertaking is obtainable from Chenvidyakarn and Woods (2006); the kind of result that will be achieved is shown in Fig. 4.8. Figure 4.8 is plotted based on a small theatre with a floor area of 64 m^2, a vent area of 0.5 m^2 and a height of 4 m. After a performance, the theatre is left at an uncomfortable temperature of 27°C when the exterior air temperature is 24°C. A comfort temperature range of 20–22.5°C is adopted, along with a required interface height of 2 m (so as to keep the hot layer above the heads of the occupants). Both of these comfort requirements are superimposed on Fig. 4.8 as the shaded zone and the dotted line, respectively. The appropriate value of G may be identified as one that allows the interface to be arrested above the desired height at the first equilibrium, and that simultaneously makes the temperature in the lower occupied zone lie within the defined comfort limits. For this example, the appropriate value of G turns out to be around 0.5. This value of G may be achieved

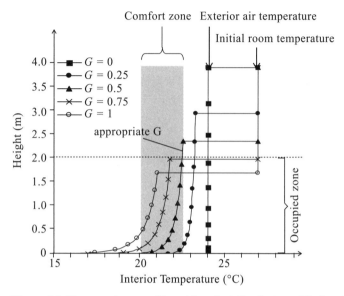

Figure 4.8. Temperature profile achieved at the first equilibrium in a small theatre flushed with pre-cooled air. (Based on calculations by Chenvidyakarn & Woods 2006.)

either by adjusting the amount of cooling or the effective vent size. (We recall from Eqs. (4.1-i)–(4.1-ii) that G is related to the rate of flow, which, in turn, is related to the effective size of the vents.) To estimate the flushing time we may use a plot similar to Fig. 4.7a. For the present example, it may be shown that the first equilibrium is reached after about 15 minutes, meaning that the performance may resume relatively quickly after each round.

As a final remark, we note that the space that we have just analysed has a stack rather than the usual flush window as the high-level opening. The reason for this is that the incorporation of the stack allows us to concentrate on the effects of cooling without having to worry about the effects of exchange flows, which may occur across the upper opening. This is because, provided that the cross-sectional area A_S of the stack and its diameter d are sufficiently small, the speed of inflow/outflow through the stack will greatly exceed the speed of any local exchange flow: $Q/A_S \gg (g'_U d)^{1/2}$. This keeps the flow through the stack uni-directional. In contrast, if a relatively wide stack or an opening flush with the envelope of the enclosure were used, exchange flows would likely develop when the ventilation were slow. This, in turn, would lead to complex evolution of the interior temperature, which, if taken into account, would make the effects of cooling difficult, or even impossible, to discern. An example of an actual building where this situation could occur is the Auden Theatre in Norfolk, UK, (Fig. 4.9). Here, possible exchange flows across its relatively short roof stacks could lead to alternations between multiple modes of ventilation, resulting in temperature fluctuations in the occupied zone, even in the absence of cooling (see Kenton, Fitzgerald & Woods 2004).

4.2. Pre-cooled ventilation of occupied spaces

The pre-cooling system discussed in Section 4.1 can also be applied to a steadily occupied space. In this case, the system does not oscillate but develops a steady flow regime. The flow regime that actually develops depends primarily on whether the space is cooled to a temperature above that of the exterior air or below it. The former cooling scheme is required if thermal comfort is to be maintained in an environment in

Figure 4.9. The Auden Theatre in Norfolk, UK.

which the ambient air temperature lies within or below the lower bound of a comfort zone, but the space contains so large a heat load that it would become uncomfortably hot without additional cooling. The latter cooling scheme, on the other hand, is appropriate when the ambient air is warmer than the upper bound of the comfort zone, as may be the case in high summer, for example.

4.2.1. Cooling to above ambient air temperature

When a space is subjected to cooling but still kept at a temperature above that of the exterior air, an upward displacement flow develops at steady state. This flow structure, first

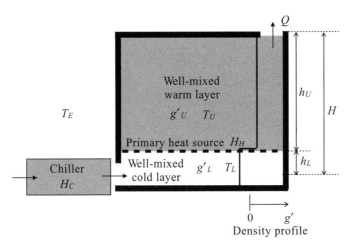

Figure 4.10. Steady upward displacement flow of a room containing a distributed source of heat and pre-cooling, when the room is kept at a temperature above that of the exterior. (Following Chenvidyakarn & Woods 2004.)

observed by Chenvidyakarn and Woods (2004), is sketched in Fig. 4.10. This sketch is based on the ventilation of a space in which the primary heat source is distributed uniformly, analogous to an audience in a fully packed theatre. The space has an undercroft that is connected to a chiller that pre-cools intake air. At steady state, net thermal buoyancy draws fresh air at a temperature below that of the ambient air into the occupied zone through the undercroft, while driving foul, warm air in the occupied zone upwards and out to the exterior through the upper vent. The system is thermally stratified into two layers. The upper layer, of depth h_U say, encompasses the main occupied zone, from the level of the primary heat source (which would be about 40 cm from the floor in a space with seated occupants and about 1 m from the floor in a space with standing occupants) to the level of the top vent. This upper layer is well-mixed, thanks to the uniform heating, and of temperature excess $\Delta T_U = T_U - T_E$ above the exterior air. The lower layer, of depth h_L say, occupies the zone from the level of the primary heat source downwards, covering the lower leg zone down to the base of the undercroft. This lower layer is also well mixed but of temperature deficiency $\Delta T_L = T_E - T_L$ below the exterior air.

At steady state, mass conservation dictates that the flux of inflow from the chiller to the undercroft, the flux of inflow from the undercroft to the room, and the flux of outflow from the room to the exterior through the upper vent are all equal, and that all represent the net volume flux. This net volume flux, denoted by Q say, may be given in terms of reduced gravity as

$$Q = A^* \left(g'_U h_U - g'_L h_L \right)^{1/2}, \qquad (4.2.1\text{-i})$$

where $g'_U = g\alpha(T_U - T_E)$ is the reduced gravity associated with warm air in the upper layer and $g'_L = g\alpha(T_E - T_L)$ is the competing reduced gravity associated with cold air in the lower layer. The variable A^*, defined along the lines of Eq. (2.4-ii), is the effective opening area taking into account pressure loss across the vents, the undercroft, the floor diffusers and the chiller. Furthermore, the global conservation of energy requires that, at steady state, heat flux H_H given out by the primary heat source balances the sum of cooling flux H_C

produced by the chiller and heat loss H_V driven by the ventilation,

$$H_H = H_C + H_V, \qquad (4.2.1\text{-ii})$$

where

$$H_C = \rho C_P Q (T_E - T_L) \qquad (4.2.1\text{-iii})$$

and

$$H_V = \rho C_P Q (T_U - T_E), \qquad (4.2.1\text{-iv})$$

with ρ the density of air and C_P the specific heat capacity of air. The flux of cooling H_C may be given by

$$H_C = \beta Q^m (T_E - T_C), \qquad (4.2.1\text{-v})$$

which is essentially the same as the relation (4.1-iii) given for the system of draining with pre-cooled air. Again, the constant β describes the properties of air and the geometry of the chiller. The exponent m describes the behaviour of flow through the chiller. It takes the value of 1/2 when the flow is laminar and 4/5 when the flow is turbulent. The chiller temperature T_C may be taken as constant for large thermal mass or a mechanical cooling device, whose temperature deviates little to none during the ventilation process. It should be, however, treated as a function of time for small thermal mass, whose temperature evolves appreciably over the period of ventilation. Combining Eqs. (4.2.1-i)–(4.2.1-v), we obtain expressions for the volume flow rate and the temperature in the upper and lower zone in terms of the heating by the internal source and the cooling by the chiller, namely

$$Q = A^{*2/3} \left(\frac{g\alpha}{\rho C_P} \right)^{1/3} [H_H h_U - \beta Q^m (T_E - T_C) H]^{1/3},$$

(4.2.1-vi)

$$T_U = T_E + \left[\frac{1}{(\rho C_P)^{2/3} (g\alpha)^{1/3} A^{*2/3}} \right]$$
$$\times \left\{ \frac{H_H - \beta Q^m (T_E - T_C)}{[H_H h_U - \beta Q^m (T_E - T_C) H]^{1/3}} \right\} \quad (4.2.1\text{-vii})$$

and

$$T_L = T_E - \left[\frac{1}{(\rho C_P)^{2/3} (g\alpha)^{1/3} A^{*2/3}} \right]$$
$$\times \left\{ \frac{\beta Q^m (T_E - T_C)}{[H_H h_U - \beta Q^m (T_E - T_C) H]^{1/3}} \right\}, \quad (4.2.1\text{-viii})$$

where $H = h_U + h_L$ denotes the whole height of the system from the mid-point of the lower opening to the mid-point of the upper opening. The terms in the square bracket of Eq. (4.2.1-vi) indicate that whereas positive buoyancy associated with the heating (of flux H_H) acts on the height h_U of the upper zone only, negative buoyancy associated with the cooling (of flux $\beta Q^m (T_E - T_C)$) acts on the whole height H of the system. It is seen later that this discrepancy in the effective heights of the two buoyancy columns contributes to a curious behaviour in the system.

As in the case of flushing ventilation with pre-cooling discussed in Section 4.1, the behaviour of the present system is controlled primarily by the strength of positive buoyancy relative to that of negative buoyancy. This relationship may be expressed in terms of parameter

$$\lambda = \frac{H_{C,ref}}{H_H} = \frac{\beta \, Q_{ref}^m \, (T_E - T_C)}{H_H}, \qquad (4.2.1\text{-ix})$$

where Q_{ref} is the reference flow rate taken as the value achieved in a room heated with a distributed source without pre-cooling (see Eq. (3.3.1-vi)). The value of λ, as we can see, represents the relative cooling input to the system.

Chenvidyakarn and Woods (2004) calculated the impact of this parameter λ on the steady state temperature in the upper and lower zones for their experimental conditions (with $m = 1/2$ and $\beta = 3170$ J/m$^{3/2}$ s$^{1/2}$ K). The results are reproduced here as the curves in Fig. 4.11. The numerical values described by these curves are, of course, specific to the experimental cases investigated, but the general trends of behaviour of the system carry through to other cases with different values of m and β. As expected, an increase in the relative strength of cooling, λ, leads to the temperature in the undercroft decreasing (Fig. 4.11a). However, the temperature response of the upper, main occupied zone is more complex (Fig. 4.11b). For small amounts of cooling, the upper zone also cools. This reduces the ventilation flow, thanks to the influence of negative buoyancy associated with the pre-cooled air. As the cooling increases, the ventilation decreases further. Eventually, the flow becomes so small that the intake air spends so much time passing through the primary heat source that the upper zone instead becomes *warmer as the cooling*

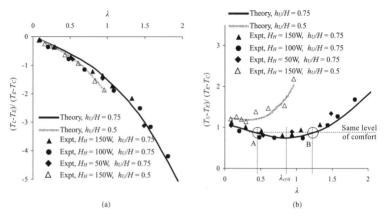

Figure 4.11. Impacts of pre-cooling on the temperature in the lower zone (a) and upper zone (b). Theoretical predictions (curves) are shown alongside the corresponding laboratory data (symbols), both of which are obtained from Chenvidyakarn and Woods (2004).

increases. This non-monotonic response of the system is rather surprising because one normally expects a space to always become colder when more cooling is supplied to it. Nonetheless, it has been confirmed experimentally, as can be seen from the symbols in Fig. 4.11b. In fact, this non-monotonic behaviour is a key characteristic that distinguishes the system of pre-cooled natural ventilation from typical air conditioning. In the latter, greater input of cooling always leads to a lower temperature, thanks to the flow rate being controlled mechanically and therefore independent of the strength of buoyancy.

It is this non-monotonic behaviour that also influences the fundamentals of how the system of pre-cooled natural ventilation should be designed and controlled. First, it can be readily seen from Fig. 4.11b that it is possible to achieve an identical temperature, and hence an identical level of thermal comfort, in the upper occupied zone using two different rates of cooling, as indicated by point A and point B. However, cooling rate A obviously requires less energy than cooling rate B. Therefore, if energy efficiency and thermal comfort are to be achieved simultaneously, cooling must be delivered at rate A. This example demonstrates how an effective control strategy needs to be developed based on a recognition of the non-monotonic behaviour of the system.

Furthermore, it can be observed that between cooling rate A and cooling rate B there is a critical cooling rate, denoted by λ_{crit}. This critical cooling rate is so called because it is the rate of cooling that produces the lowest temperature possible in the upper zone; any cooling input greater than this will result only in the room heating up rather than cooling down. Thus, this critical cooling rate effectively defines the cooling capability of the system. Its value, which is identifiable by differentiating the temperature in the upper zone with respect to the flow rate and equating the result with zero, is sensitive to the vertical position of the primary heat source relative to the vertical separation between the vents, h_U/H. It increases as the primary heat source is moved downwards or if the upper opening is moved upwards (so that the value of h_U/H increases; see Fig. 4.12a), reflecting the fact that the taller the occupied zone is, the greater positive buoyancy drive there is, and the more difficult it is for the flow to stall. For a space with a typical floor-to-ceiling height of 3 m and a 50 cm

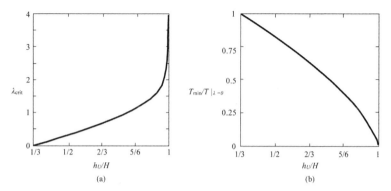

Figure 4.12. Impacts of the relative position of the primary heat source, h_U/H, on the critical cooling rate λ_{crit} and the minimum temperature achievable in the upper layer, T_{min}. The plot is based on $m = 1/2$, but the general trend applies also to the case where $m = 4/5$. (Based on model by Chenvidyakarn & Woods 2004.)

deep undercroft ($h_U/H = 5/6$), this critical cooling rate may come in at about 1.2 times the size of the heat load in the space, depending on the effectiveness of the chiller. This critical cooling rate will increase to about 1.4 times the size of the heat load if the height of the room is increased to 4.5 m. Note, for a given heat load and chiller, there is usually a critical depth of the upper zone such that if the upper zone is any shallower than this, ventilation will become so suppressed that cooling only heats up the upper zone (i.e., $\lambda_{crit} = 0$) and never cools it. For the preceding example, this critical depth of the upper zone is one-thirds of the overall height of the system, $h_U/H = 1/3$.

   To determine the lowest temperature achieved at the critical cooling rate we may combine Eq. (4.2.1-vii) with

Eq. (4.2.1-ix). Obviously, this lowest temperature is also sensitive to the relative depth of the upper zone. See Fig. 4.12b. For large h_U/H, $h_U/H \to 1$, corresponding to having the top vent located high above the level of the primary heat source, there is rapid flow flushing a large amount of heat out of the room, and so the minimum temperature in the main occupied zone is low. For small h_U/H, corresponding to having the top vent located close to the level of the primary heat source, there is relatively slow flow and hence small flushing of heat. As a result, the minimum temperature in the upper zone is high. For a space with a typical floor-to-ceiling height of 3 m and a 50 cm deep undercroft mentioned earlier, the lowest temperature achieved at the critical cooling rate is about 65% below the temperature that would be achieved if there were no cooling at all ($\lambda = 0$). As the depth of the occupied zone is increased from 3 m to 4.5 m, a decrease in the lowest temperature by about 30% is effected.

At this point it is important to also note that, owing to the stratification in the space, there is a possibility that the occupants may experience a cold feet sensation. This is caused by the temperature in the lower layer falling below a comfort value. The problem, however, may be prevented by carefully matching the rate of cooling with the size of the internal heat load, taking into account the target ventilation rate and comfort temperature range. In principle, the appropriate rate of cooling must be sufficiently large to allow effective cooling of the upper body zone, while moderate enough not to cause draughts in the lower leg zone or insufficient ventilation. For a given size of heat load, there is usually a range of cooling rates that can satisfy these conditions. These cooling rates are usually fractions of the heat load in the space. For example, an auditorium with a height of 10 m, a 3 m deep undercroft and an effective opening area of about 1 m^2 may require a cooling flux between 30% and 50% of the heat load to accommodate about 90 people, depending on the effectiveness of the chiller and the comfort standards adopted. Smaller cooling requirement may be expected if the space has a larger vent or a taller upper zone, owing to the enhancement of heat removal through increased ventilation.

As a side remark, we note that the system of pre-cooling described in this section can also be adapted to provide pre-heating. In fact, it is already quite common for cooling mediums such as a cooling coil in a heat pump system to be used as a heating coil on colder days, or a rock bed to be used for storing heat in winter as well as for providing cooling in summer. This allows the system to be more versatile, making it usable in both summer and winter. However, the temperature structure that develops in a space equipped with pre-heating is quite different from that which develops in a space equipped with pre-cooling. With pre-heating, the upper occupied zone attains a temperature above that of the undercroft as before, but the undercroft now attains a temperature *above* that of the exterior air rather than below it. Thus positive buoyancies in the two parts of the space combine to drive the ventilation rather than compete with each other. To reflect this, some sign changes have to be made in the model given above. If we denote the heater temperature by T_H and the flux of pre-heating by H_{PH}, we may rewrite Eqs. (4.2.1-v)–(4.2.1-viii) respectively as

$$H_{PH} = \beta Q^m (T_H - T_E) \qquad (4.2.1\text{-x})$$

$$Q = A^{*2/3} \left(\frac{g\alpha}{\rho C_P} \right)^{1/3} [H_H h_U + \beta Q^m (T_H - T_E) H]^{1/3},$$
$$(4.2.1\text{-xi})$$

$$T_U = T_E + \left[\frac{1}{(\rho C_P)^{2/3} (g\alpha)^{1/3} A^{*2/3}} \right]$$
$$\times \left\{ \frac{H_H + \beta Q^m (T_H - T_E)}{[H_H h_U + \beta Q^m (T_H - T_E) H]^{1/3}} \right\} \qquad (4.2.1\text{-xii})$$

and

$$T_L = T_E + \left[\frac{1}{(\rho C_P)^{2/3} (g\alpha)^{1/3} A^{*2/3}} \right]$$
$$\times \left\{ \frac{\beta Q^m (T_H - T_E)}{[H_H h_U + \beta Q^m (T_H - T_E) H]^{1/3}} \right\}. \qquad (4.2.1\text{-xiii})$$

Obviously, with no competition between positive and negative buoyancy, the system equipped with pre-heating does not exhibit a non-monotonic behaviour. Therefore, the whole system simply warms up as the rate of heating is increased. For a given size of heat load, the appropriate range of heating rates is one that is sufficiently large to warm the lower leg zone but not so great that it overheats the upper body zone.

4.2.2. Cooling to below ambient air temperature

Quite a different control strategy from that described in Section 4.2.1 will be required if the room is cooled to a temperature below that of the exterior air. The flow regime in this case is, broadly speaking, an invert of that developed in the room cooled to a temperature above the exterior air discussed earlier: with the interior air now being heavier than the ambient air, the flow is directed downwards, and the room draws fresh air through its upper opening while venting through its lower opening. To allow the occupants located on the floor to be cooled effectively in this flow regime, pre-cooled air needs to be introduced at a high level instead of at a low level. To this end, a ceiling space may be used as an air flow channel to distribute pre-cooled air through ceiling diffusers, as sketched in Fig. 4.13, for example.

The mechanics of 'top-down pre-cooled natural ventilation', as the system may be called, have been examined by Chenvidyakarn and Woods (2005b). They showed that if the primary heat source is uniformly distributed, the room becomes well-mixed at steady state. At this state, mass conservation dictates that the inflow volume flux through the chiller, the volume flux through the ceiling diffusers and the outflow volume flux through the lower opening are all equal, and that all represent the net ventilation flow. This net ventilation flow is driven by the reduced gravity $g' = g\alpha(T_E - T_{IN})$ associated with temperature deficiency in the space; that is,

$$Q = A^* \, (g'H)^{1/2}, \qquad (4.2.2\text{-i})$$

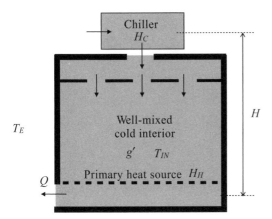

Figure 4.13. Steady downward displacement flow of a room containing a distributed source of heat and pre-cooling, when the room is cooled below the temperature of the exterior air. (Following Chenvidyakarn & Woods 2005b.)

where A^*, given according to Eq. (2.4-ii), is the effective opening area taking into account pressure loss across the chiller, the ceiling space, the ceiling diffusers and the openings of the room, and H is the vertical separation between the upper and lower openings. Furthermore, the conservation of thermal energy requires that heat removal by the chiller, H_C, balances the sum of heat gain H_V from the ventilation, and heat gain H_H from the internal source,

$$H_C = H_V + H_H, \qquad \text{(4.2.2-ii)}$$

where

$$H_V = \rho C_P Q (T_E - T_{IN}) \qquad \text{(4.2.2-iii)}$$

and H_C is given by Eq. (4.2.1-v). Combining this last equation with Eqs. (4.2.2-i)–(4.2.2-iii), we obtain the volume flow rate and the interior temperature at steady state in terms of the flux of heating by the internal source and the flux of cooling by the chiller, namely

$$Q = A^{*2/3} \left(\frac{g\alpha}{\rho C_P} \right)^{1/3} [\beta Q^m (T_E - T_C) - H_H]^{1/3} H^{1/3} \quad \text{(4.2.2-iv)}$$

and

$$T_{IN} = T_E - \frac{[\beta Q^m (T_E - T_C) - H_H]^{2/3}}{H^{1/3} (\rho C_P)^{2/3} (g\alpha)^{1/3} A^{*2/3}}. \quad \text{(4.2.2-v)}$$

Equation (4.2.2-iv) shows that, unlike in the case of upward pre-cooling discussed in Section 4.2.1, where the effective heights of the positive and negative buoyancy columns are different, leading to a critical height of the primary heat source above which cooling becomes ineffective, in the case of downward pre-cooling the heights of the positive and negative buoyancy columns are identical: both types of buoyancy act over the whole height H of the system. Consequently, the system responds monotonically to an increase in cooling: the temperature in the space always decreases as the cooling is increased (Fig. 4.14). This is in contrast to the case of upward pre-cooling, in which excessive cooling can inadvertently warm the space.

However, as with the case of upward pre-cooling, the behaviour of the system with downward pre-cooling is governed by the strength of cooling relative to the strength of heating. In fact, the stability of the downward flow regime depends on the negative buoyancy being sufficiently stronger than the positive buoyancy (Fig. 4.15). For a given chiller and a given geometrical configuration of the space, there

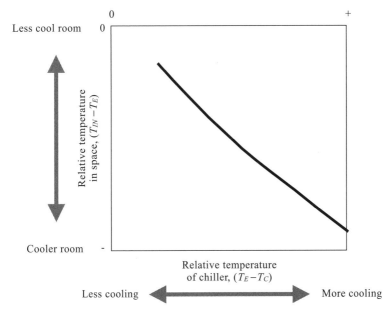

Figure 4.14. Temperature response of a room equipped with top-down pre-cooling. (Based on model by Chenvidyakarn & Woods 2005b.)

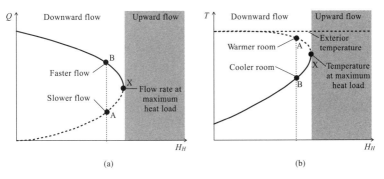

Figure 4.15. Stable and unstable flow regimes associated with downward pre-cooled natural ventilation. (Based on model by Chenvidyakarn & Woods 2005b.)

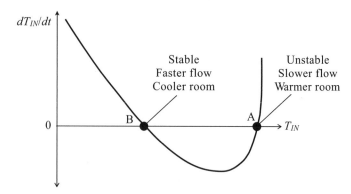

Figure 4.16. Response of a system of downward pre-cooled natural ventilation to a small perturbation. (Based on model by Chenvidyakarn & Woods 2005b.)

is a maximum heat load (indicated by point X in Fig. 4.15) within which a downward flow can be stably maintained, and above which an upward flow develops instead. For smaller heat loads, Eq. (4.2.2-iv) and Eq. (4.2.2-v) admit two solutions: a slow flow with an interior temperature close to that of the exterior air (represented by point A in Fig. 4.15) and a fast flow with an interior temperature much lower (represented by point B in Fig. 4.15). However, the slow flow regime is unstable as may be recognised by considering the impact of a small perturbation to this regime. See Fig. 4.16. When the system is out of equilibrium but well-mixed, the interior temperature evolves according to the balance between the heat input to the system and the heat removal from it, namely

$$\rho C_P V \frac{dT_{IN}}{dt} = H_H + \rho C_P Q (T_E - T_{IN}) - \beta Q^m (T_E - T_C).$$

$$(4.2.2\text{-vi})$$

If the system is initially at equilibrium in the slow regime (point A), a small perturbation caused, for example, by a gust of wind or the opening of a window will lead to a sudden increase in the volume flow and hence a small decrease in temperature in the space. This, in turn, will lead to a greater influx of pre-cooled air, which reduces the interior temperature even further. As a result, the system will settle to a cooler equilibrium sustained by a faster flow (point B). At this point, if the same kind of perturbation occurs, it will again lead to an increase in the volume flow and a small decrease in temperature in the space. However, the flow through the chiller will be so fast that the intake air does not have much time to be cooled by the chiller. Therefore, the flux of cooling delivered to the room will reduce, resulting in the system warming up and adjusting back to the equilibrium point B. If, at this point, there is instead temporary closing of a window, there will be a small reduction in the ventilation flow and a small rise in temperature in the space. The slowing of the flow will increase the amount of time the intake air spends travelling through the chiller so that it is cooled to a lower temperature than before. Therefore, the cooling flux delivered to the room increases. Consequently, the system cools down and adjusts back to the equilibrium point B.

The maximum heat load that allows a stable downward flow to be maintained can be determined from differentiating the heat load H_H with respect to the flow rate Q and equating the result with zero (consider Eq. (4.2.2-iv) and Fig. 4.15a). This maximum heat load is proportional to the downward driving force, and so increases with the vent separation H, the effective vent area A^*, the heat transfer effectiveness of the chiller, β and m, and the difference between the temperature of the chiller and the temperature of the exterior air $(T_E - T_C)$. For example, Chenvidyakarn and Woods (2005b) showed that for a system equipped with a flat plate chiller with $m = 1/2$, the maximum heat load is given by $H_{H\text{max}} = (5/6^{6/5})[\beta(T_E - T_C)]^{6/5}(A^{*2}g\alpha H)^{1/5}(\rho C_P)^{-1/5}$.

An important remark must be made at this point: the maximum heat load that allows a stable downward flow to be maintained is *not* necessarily the maximum heat load that the system can take while delivering sufficient ventilation, thermal comfort and energy efficiency simultaneously. In fact, the latter heat load is usually lower than the former; and to satisfy all the key goals of ventilation, the heat load must be carefully matched with the vent size and the cooling rate. This control principle may be illustrated using a simple scenario. Consider an auditorium or a meeting hall whose number of occupants (and hence heat load H_H) varies from 50 to 300, depending on its use. To achieve thermal comfort, sufficient ventilation and energy efficiency simultaneously in this auditorium, its occupancy pattern must obviously be taken into account. Suppose, for the sake of discussion, that the vents of the space can be adjusted to three different discrete sizes, namely 2.5, 5 and 7 m^2, so that, assuming certain geometrical configurations and settings for the system, the aforementioned model may produce a plot similar to Fig. 4.17. (In practice, the opening size may be adjusted smoothly or discretely depending on the control mechanism, but this does not affect our present argument.) The plot shows, as a point on

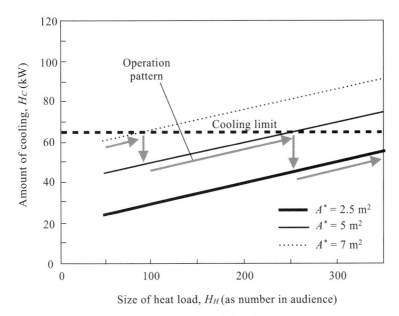

Figure 4.17. Coordinated adjustment of the effective vent size and the amount of cooling in a room equipped with downward pre-cooling. (Based on model by Chenvidyakarn & Woods 2005b.)

each line, the amount of cooling required to achieve thermal comfort and sufficient ventilation for a given size of heat load at a given size of vent. Every point on every line represents a rate of ventilation greater than a minimum requirement and an identical temperature required to achieve thermal comfort. It can be seen that if the vent is fixed at a certain size (i.e., a certain line), the amount of cooling must always be increased if larger heat loads are to be accommodated. For instance, if the vent size is held at 7 m^2, to increase the capacity of the space from 50 to 100 people, the amount of cooling has to be increased from about 60 kW to 65 kW, and to accommodate 250 people it has to be increased to about 80 kW.

Obviously, increasing the amount of cooling alone is not an energy-efficient operation strategy, and there could also be some limit imposed on the maximum amount of cooling that is allowed to be reached (represented by the horizontal dashed line in Fig. 4.17), which stems from either the desire to control energy consumption or the capacity of the chiller itself. In this example, if the limit of the chiller is placed at say 65 kW, then increasing the cooling flux alone without adjusting the vent size from $A^* = 7$ m^2 will never allow the auditorium to reach its full capacity of 300 people while maintaining a comfortable internal temperature; the chiller capacity will always limit the number in the audience to just about 100 people.

A more effective, energy-efficient operation strategy may be adopted which is based on coordinated adjustment of the cooling rate and the effective vent size to match the varying heat load. The underlying principle of this is that the same size of heat load can be accommodated with less cooling at smaller vents, because larger vents allow more ventilation, thereby requiring the chiller to cool a larger amount of air passing through it. Adopting this principle, a pattern of operation akin to that indicated by the grey arrows may be developed, whereby the vent size is reduced when the cooling amount reaches a predefined limit. For instance, with the vent size at 7 m^2, to accommodate 100 people, a cooling flux of about 65 kW – the limit of the chiller – must be supplied. At this point, if the vent size is reduced to 5 m^2, it will be possible to accommodate the same number of people using a smaller amount of cooling: only about 50 kW, in fact. To increase the capacity of the auditorium at this new vent size, the cooling flux may be increased until the limit of the chiller is reached again, at which point about 250 people may be accommodated. To allow a greater number of people, the vent size must be reduced to 2.5 m^2 and the cooling increased accordingly. Clearly, this operational strategy not only helps extend the capability of the system, but is also conducive to energy efficiency. Note, in many circumstances, just like in the preceding example, it may be undesirable to always keep the vents at the smallest size possible, even though this will obviously lead to the greatest energy conservation. This is because there may be a need to optimise the flushing of contaminants from the space to achieve the best indoor air quality possible.

4.3. Maintained source of heat and internal cooling

Pre-cooling of intake air using a medium located outside the ventilated space is but one way of providing cooling; a reduction in the interior temperature can also be accomplished through a cooling device located within the space itself. Such a device may take the form of a mechanical chilled ceiling or a chilled beam, or thermal mass that is part of the structure of the building, such as a concrete or stone floor slab or ceiling.

To be effective in providing cooling, the surface of the chilled ceiling/thermal mass must be uninsulated and exposed to the interior space to allow good heat exchange between the ceiling/mass and the air in the space. This is usually not a problem in the case of an industrially produced chilled ceiling, since it is normally designed from the factory to hide unsightly mechanical/electrical works and to display its well-treated surface which immediately fulfils aesthetic requirements. However, in the case of a thermal mass ceiling that forms part of the structure of the building, this is often not the case. The surface of such a ceiling is often rough and space is not often dedicated above or within it for mechanical/electrical works. Therefore, a false ceiling is often required to be fitted below it for aesthetic purposes. This false ceiling, usually made of a lightweight material such as gypsum boards, prevents an effective coupling between the mass and the space. In order that the surface of the mass could be exposed, decisions would normally need to be made from the beginning of the design process to do so, so that alternative construction could be pursued. Examples of buildings in which this is the case include the Building Research Establishment's low energy office in Watford and the Millennium Galleries in Sheffield, both in the United Kingdom. In the former, exposed sinusoidal shaped concrete ceilings are used to provide passive cooling for the main working floors (Fig. 4.18; see Crisp, Fisk & Salvidge 1984). In the latter, bare yet well-finished concrete floor slabs are used for similar purposes in the back-of-house area (see Long 2001). In both cases (as in most others similar to them), the choice of how to incorporate thermal mass depends not only on the thermal capacity required, but also on other practical architectural and engineering considerations such as look and weight.

In modelling the behaviour of an internal source of cooling, different approaches may be used, depending primarily on the cooling mechanism of the source itself. If the source is mechanical in nature, such as a chilled ceiling, the

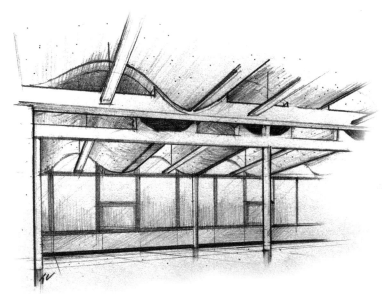

Figure 4.18. The sinusoidal-shaped concrete ceiling of the Building Research Establishment building, Watford, UK.

procedure is relatively straightforward because the temperature of the ceiling may be arbitrarily controlled. In this case, the ceiling temperature may be assumed constant or prescribed to follow a specific pattern during the ventilation period. In contrast, the behaviour of a thermal mass ceiling/floor is more complex because its temperature evolves over time in response to changes in the exterior air temperature, the internal heat load and the ventilation rate. Approaches for modelling this thermal evolution process are described in relative detail by Holford and Woods (2007) and Holman (1997), and are touched on in many good textbooks on heat transfer. These are quite involved procedures and beyond the scope of this book. For the present purposes, it suffices to note that internally exposed thermal mass typically has a heat capacity 10–100 times that of air, owing to its much greater density. Consequently, the timescale over which the temperature of the mass evolves is also 10–100 times as long as that of air (Fisk 1981; Li & Yam 2004). This temporal discrepancy leads to a phase lag between the temperature response of the mass and the temperature response of the air in the interior of a building, which in turn leads to appreciable convective heat exchange between the two elements. In typical buildings with ventilation rates of about three to four air changes per hour, natural ventilation flows reach quasi-steady state within tens of minutes. This is in contrast to thermal mass, which typically takes hours. Thus, provided that the mass is sufficiently large, we may assume that the temperature of the mass remains constant for a sufficiently long period of time that a quasi-steady-state ventilation regime is allowed to develop, but that over a longer timescale the temperature of the mass evolves substantially so that the quasi-steady-state ventilation regime changes. This allows us to model the convective heat transfer associated with the ventilation of the building as driven by a combination of convective sources associated with the occupancy of the building and heat sinks associated with the mass.

The aforementioned approach is taken in our examination below of the basic flows driven by a combination of an internal source of cooling and an internal source of heat. Our examination begins with a space in which both the source of cooling and the source of heating are distributed, as in,

Figure 4.19. The design studio at the University of Cambridge Department of Architecture.

for example, a uniformly occupied open-plan office equipped with a chilled ceiling. We then continue with a space where a localised source of heat is present alongside a distributed source of cooling, as may be the case when the aforementioned office is locally occupied. Finally, a problem is considered in which both the source of cooling and the source of heating are localised, as may be found in an atrium occupied in one spot whose roof contains a concentrated area of poorly insulated glass, for instance.

4.3.1. Distributed source of heat and distributed source of cooling

The design studio at the University of Cambridge Department of Architecture (Fig. 4.19) is an example of a space containing a combination of a distributed source of heat and a distributed source of cooling. Here, the students spreading out across

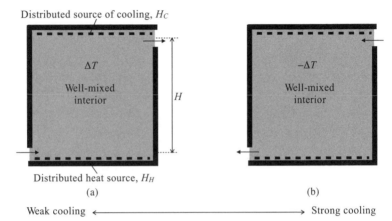

Figure 4.20. Steady flow regimes in a room cooled by a distributed source at the ceiling and heated by a distributed source at the base, when the heating dominates (a) and when the cooling dominates (b).

the studio act as a source of heat while the sawtooth-shaped chilled ceiling provides cooling; windows located at high and low levels facilitate natural ventilation. To understand the principles of what goes on in this space and those of similar set-ups, it is convenient to use the simplified diagram in Fig. 4.20. In this diagram, the students are treated as a uniform areal source of heat located at the base of the space, and the sawtooth-shaped ceiling is treated as flat. Furthermore, the openings are assumed to be vertically short compared to the vertical extent of the space so that a uni-directional flow develops across each opening. These simplifications may be made without affecting the basic overall flow picture.

Observation of flows developed in this space using water-bath experiments reveals that warm convective plumes from the heat source at the base of the space rise as cold convective plumes from the ceiling descend. These warm and cold plumes mix the air in the space while providing buoyancy, driving natural ventilation. In these conditions, two ventilation regimes are possible, depending on the relative strength between the cooling and the heating. When the magnitude of heating is larger than that of cooling, the room converges to a steady state in which an upward displacement flow maintains a well-mixed interior at a temperature above that of the ambient air, as depicted in Fig. 4.20a. This flow structure, we may immediately recall, is not dissimilar to what we have seen in the upper occupied zone of a space ventilated with upward pre-cooling discussed in Section 4.2.1. At steady state, mass conservation requires that the inflow volume flux through the lower opening matches the outflow volume flux through the upper opening. Each of these volume fluxes represents the net volume flow. It is driven by the reduced gravity $g' = g\alpha\Delta T = g\alpha(T_{IN} - T_E)$ associated with temperature excess in the space,

$$Q = A^* (g\alpha\Delta TH)^{1/2}, \qquad (4.3.1\text{-i})$$

where A^* is the effective opening area, H is the vertical separation between the upper and lower openings, g is gravitational acceleration and α is the volume expansion coefficient of air. In addition, the global conservation of thermal energy requires that heat gain from the distributed source of flux H_H balances heat removal by the ceiling of flux H_C and heat loss driven by the displacement ventilation of flux H_V:

$$H_H = H_C + H_V, \qquad (4.3.1\text{-ii})$$

where

$$H_V = \rho C_P Q \Delta T, \qquad (4.3.1\text{-iii})$$

with ρ being the density of air and C_P the specific heat capacity of air.

For the case in which the magnitude of cooling is sufficiently stronger than that of heating, we may deduce from our discussion in Section 4.2.2 that a steady downward displacement flow develops instead of an upward one. This flow structure maintains the room at a temperature below that of the exterior air, as sketched in Fig. 4.20b. Equations (4.3.1-i)–(4.3.1-iii) still apply in this case, but the values of ΔT and Q are now negative, reflecting the temperature deficiency in the space and the downward direction of flow, respectively.

In both flow regimes, the response of the system to variation in the magnitude of cooling depends greatly on how the cooling is provided. If a mechanically chilled ceiling is used, as in the actual studio building, the cooling flux may be independently adjusted to correspond to the heat load (see Novoselac & Srebric 2002). Therefore, H_C may be treated as an independent variable. In this case, combining Eqs. (4.3.1-i)–(4.3.1-iii) gives the net volume flux and interior temperature in terms of the flux of cooling and the flux of heating as

$$Q = A^{*2/3} \left(\frac{g\alpha}{\rho C_P} \right)^{1/3} (H_H - H_C)^{1/3} \, H^{1/3}, \quad (4.3.1\text{-iv})$$

and

$$T_{IN} = T_E + \frac{(H_H - H_C)^{2/3}}{(\rho C_P)^{2/3} (g\alpha)^{1/3} A^{*2/3} H^{1/3}}. \quad (4.3.1\text{-v})$$

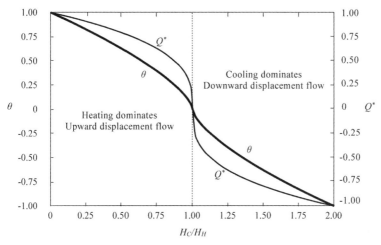

Figure 4.21. Impacts of the relative strength of cooling, H_C/H_H, on the volume flow rate and interior temperature of a room cooled by a distributed source at the ceiling and heated by a distributed source at the base, when the cooling rate can be independently adjusted. The flow rate and interior temperature are shown in dimensionless terms as scaled on those of a room without cooling (see Section 3.3.1), as Q^* and θ respectively, where $Q^* = Q/Q_0 = Q/\{A^{*2/3}[g\alpha/(\rho C_P)]^{1/3}H^{1/3}H_H^{1/3}\}$ and $\theta = \Delta T/\Delta T_0 = \Delta T (A^*\rho C_P)^{2/3}(g\alpha H)^{1/3}/H_H^{2/3}$. (Based on models by Chenvidyakarn & Woods 2005b and Livermore & Woods 2008.)

Results from Eqs. (4.3.1-iv) and (4.3.1-v) are plotted non-dimensionally in Fig. 4.21 as relative to the values achieved in a room with no cooling. We can see that when the heating dominates, $H_C/H_H < 1$, and the flow is upward, an increase in the relative strength of cooling, $H_C/H_H \to 1$, leads to a reduction in temperature in the space. This, in turn, increases negative buoyancy in the system, slowing down the flow. However, when the cooling dominates, $H_C/H_H > 1$, and the flow is downward, the system responds differently. Although in this case an increase in the relative magnitude of cooling still leads to a decrease in temperature in the space, the resultant increase in negative buoyancy strengthens the flow instead of suppressing it (i.e., the flow rate is of increasing negative values).

The response of the system will be quite different from this picture, however, if the cooling is provided by a thermally massive ceiling. In this case, the temperature of the surface of the ceiling evolves over time according to convective heat exchange between the ceiling surface and the air in the room which, in turn, depends on the balance between conductive heat transfer through the ceiling material and radiative heat transfer at the ceiling surface. For this reason, H_C cannot be treated as an independent variable, and to predict the evolution of the ceiling temperature, modelling of the relevant transient heat transfer processes is required (see Holford & Woods 2007 for more details on this). An examination of the change in the quasi-steady-state interior temperature structure as the temperature of the ceiling evolves reveals the following. If the cooling by the ceiling is initially sufficiently weak compared with the heating in the room, the room vents upwards, as shown in Fig. 4.20a. As the temperature of the ceiling surface increases through various heat gains, the cooling flux from the ceiling decreases. For a room with a constant heat load, this leads to an increase in temperature in the room. In this way, the cooling capability of the system, and thus the level of thermal comfort, reduces over time. Eventually, thermal comfort may not be maintained because the temperature of the ceiling rises excessively. To rejuvenate the system, cool exterior air may be used to flush the ceiling at night after the occupancy hours to bring its temperature below that required in the room during occupancy; this technique is called *night cooling* or *purging*. If the night-time temperature of the exterior air is sufficiently low, the cooling provided by the newly rejuvenated ceiling may dominate the heating by the internal source during the early hours of the following day. This will lead to the system venting downwards initially, as sketched in Fig. 4.20b. Depending on the amount of heat load in the space and the exterior air temperature at this point, the occupants may feel uncomfortably cold. However, as the ceiling heats up by a combination of various gains, the level of thermal comfort will increase. Eventually, the cooling flux of the ceiling may be reduced sufficiently that it becomes dominated by the heating from the internal source again. As this occurs, the flow will resume its upward direction depicted in Fig. 4.20a.

4.3.2. Localised source of heat and distributed source of cooling

If the space in Section 4.3.1 contains a localised heat source of flux H_P at the base instead of a distributed heat source (e.g., when there is a concentrated group of students working in the design studio of Fig. 4.19), the picture of flow becomes richer, as reported by Livermore and Woods (2008; Fig. 4.22). Provided that the localised heat source is sufficiently strong, a plume rising from it forms a buoyant layer at a height h above the base of the room, leaving the lower part of the room filled with fresh air, akin to what was observed by Linden,

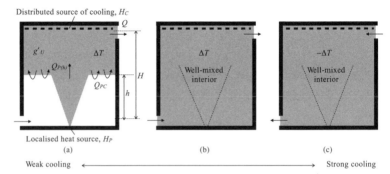

Figure 4.22. Steady flow regimes in a room cooled by a distributed source at the ceiling and heated by a localised source at the base, when the heating dominates and the cooling is relatively weak (a), when the heating dominates and the cooling is relatively strong (b), and when the cooling dominates (c). ((a) and (b) follow Livermore & Woods 2008; (c) follows Chenvidyakarn & Woods 2005b.)

Lane-Serff and Smeed (1990) in a room heated solely by a localised source at the base (see Section 3.2.3). However, with the ceiling now acting as a source of cooling, negatively buoyant convective plumes generated by the ceiling also descend within the upper layer while mixing the air therein. If these cold plumes possess sufficient momentum they will be able to travel past the interface between the upper and lower layers, penetrate into the lower layer and entrain fresh air back into the upper layer. This penetrative convection will affect the mass and thermal balances of the system, modifying the depth of the upper layer and its temperature.

Clearly, in this flow regime, whether or not the occupants in the lower part of the building can enjoy thermal comfort depends chiefly on the following factors, namely the position of the interface relative to the level of the occupied zone, the temperature within the upper buoyant layer relative to the adopted comfort standard, and the temperature of the ambient air in the lower zone relative to the comfort standard. If the interface is established above the occupied zone, the level of comfort experienced by the occupants depends on the severity of the temperature of the ambient air. If the interface is established within the occupied zone, the level of comfort depends on the magnitude of cooling relative to the size of the heat load, which controls the temperature in the buoyant layer. To develop an effective control strategy for the system, therefore, an understanding is required of how the source of cooling and the source of heating interact and affect the temperature structure in the space. Consideration of the mathematics underpinning the flow will help shed light on this.

Let us assume that a quasi-steady state is reached. At this state, mass conservation requires that the outflow volume flux Q through the upper opening matches the sum of the volume flux $Q_{P(h)}$ supplied by the plume at the height h of the interface and the volume flux Q_{PC} entrained by the penetrative convection:

$$Q = Q_{P(h)} + Q_{PC}, \qquad (4.3.2\text{-i})$$

where each volume flux is defined as follows. The volume flux Q through the upper opening is driven by the reduced gravity $g'_U = g\alpha\Delta T$ acting over the upper buoyant layer of temperature excess ΔT above the exterior; that is,

$$Q = A^* \left[g'_U (H - h) \right]^{1/2} . \qquad (4.3.2\text{-ii})$$

where H is the distance between the upper and lower openings, and $(H-h)$ is the depth of the upper layer. The volume flux supplied by the plume at the height of the interface is given according to the plume theory (Eq. (3.2.1-xiii)) as

$$Q_{P(h)} = \lambda B_P^{1/3} h^{5/3}, \qquad (4.3.2\text{-iii})$$

where λ is an integrated constant describing the entrainment of the plume, and B_P is the buoyancy flux of the localised source. The latter may be given in terms of the heat flux H_P of the source according to Eq. (3.2.1-xiv) as

$$B_P = \frac{g\alpha H_P}{\rho C_P}. \qquad (4.3.2\text{-iv})$$

The volume flux of the penetrative convection may be quantified using the approach due to Zilintikevich (1991), which simply treats the penetrative convection as bringing to the lower layer a fraction k of the buoyancy flux B_C associated with the cold ceiling that drives entrainment across the interface; that is,

$$Q_{PC} = \frac{k B_C}{g_U'}, \qquad (4.3.2\text{-v})$$

where B_C is defined along the same lines as Eq. (4.3.2-iv) in terms of the heat flux H_C of the ceiling; that is,

$$B_C = \frac{g\alpha H_C}{\rho C_P}. \qquad (4.3.2\text{-vi})$$

Finally, the temperature excess in the upper layer is given as
a function of heat gain from the plume, H_P, and heat loss
through the ceiling, H_C, as

$$\Delta T = \frac{H_P - H_C}{\rho C_P Q}. \qquad (4.3.2\text{-vii})$$

Combining Eqs. (4.3.2-i)–(4.3.2-vii), we obtain for the dimensionless interface height

$$\frac{h}{H} = \left(\frac{A^{*2}}{H^4 \lambda^3}\right)^{1/5} \left(1 - \frac{h}{H}\right)^{1/5} \frac{\left[1 - \frac{H_C}{H_P}(1 + k)\right]^{3/5}}{\left(1 - \frac{H_C}{H_P}\right)^{2/5}}, \qquad (4.3.2\text{-viii})$$

and the dimensionless temperature in the upper layer

$$\theta = \left[\frac{\left(1 - \frac{H_C}{H_P}\right)^2}{1 - \frac{h}{H}}\right]^{1/3}, \qquad (4.3.2\text{-ix})$$

where θ is scaled on the temperature of a room ventilated naturally without cooling as $\theta = \Delta T / \Delta T_0 = \Delta T$
$(A^* \rho C_P)^{2/3}(g\alpha H)^{1/3}/H_H^{2/3}$ (see Section 3.3.1). It may be noted
that with the room containing a distributed source of cooling
at the ceiling as well as a localised source of heating at the
base, the height of the interface is determined not only by
the geometrical relations of the space but also by the relative
strength of buoyancy (Eq. (4.3.2-viii)). This is in contrast to
the case of a room heated by a localised source with no cooling
discussed in Section 3.2.3, where the height of the interface is
controlled by the geometry of the space but is independent of
the buoyancy flux.

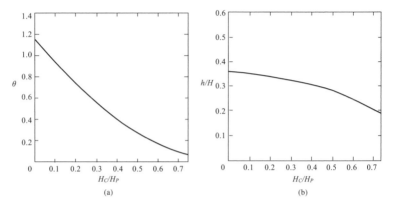

Figure 4.23. Impacts of the relative strength of cooling, H_C/H_P, on the temperature in the upper layer (a) and the height of the interface between the upper and lower layers (b) in a room cooled with a distributed source at the ceiling and heated with a localised source at the base, when the cooling rate can be independently adjusted. In this plot, $A^{*2}/(H^4\lambda^3) = 0.01$ is taken. (Based on calculations by Livermore & Woods 2008.)

As with the case of distributed heating discussed in Section 4.3.1, how best to control the system with localised heating depends strongly on the mechanism of cooling, which governs the response of the system. If the cooling is provided mechanically, for example, through a chilled ceiling, the cooling flux may be arbitrarily adjusted, and therefore H_C in Eqs. (4.3.2-vi)–(4.3.2-ix) may be treated as an independent variable. In this case, the response of the system follows the curves shown in Fig. 4.23. When the relative cooling flux $H_C/H_P = 0$, corresponding to the case where there is no cooling at all, the temperature in the upper layer and the height of the interface are as predicted by the classic plume model due to Linden, Lane-Serff and Smeed (1990;

Eqs. (3.2.3-ii)–(3.2.3-iii)). As the cooling increases so that $H_C/H_P \rightarrow 1/(1+k)$, the room cools (Fig. 4.23a) while the upper layer deepens downwards (Fig. 4.23b), causing the room to approach well-mixed conditions. Eventually, as the cooling is increased sufficiently, the room becomes well-mixed, $h/H = 0$, while the overall flow is maintained upwards (Fig. 4.22b). In these conditions, the flow rate is given by the reduced gravity acting over the entire height of the room, namely

$$Q = A^* \left(g'_U H\right)^{1/2} = A^* \left(g\alpha \Delta T H\right)^{1/2}, \qquad \text{(4.3.2-x)}$$

which is the same as Eq. (4.3.1-i) given for the room equipped with a distributed heat source and a chilled ceiling, thanks to the similarly well-mixed interior. With the room being now well-mixed, any increase in the magnitude of cooling will cause the temperature of the whole room to reduce progressively. The mean interior temperature in these conditions may be found using Eq. (4.3.2-vii). Ultimately, once the cooling flux is increased beyond a critical value, the cooling becomes dominant over the heating. At this point, we may deduce from the work on top-down pre-cooling by Chenvidyakarn and Woods (2005b; see Section 4.2.2) that the system will overturn and vent downwards instead. Furthermore, the temperature in the space will be maintained below that of the exterior air instead of above it. The flow picture is now as depicted in Fig. 4.22c. In this regime, Eqs. (4.3.1-i)–(4.3.1-v) given for the case of distributed heating are applicable, with H_H being replaced by H_P, and the flow rate and interior temperature now taking negative values to reflect the downward direction of flow and the temperature deficiency of the space relative to outside. The system thus responds to variation in the rate of cooling according to the curves in Fig. 4.21 in the limit $H_C/H_P = H_C/H_H > 1$.

As expected, the behaviour of the system will be quite different if cooling is provided instead by thermal mass, for example, through an exposed concrete ceiling. In this case, the cooling input cannot be controlled arbitrarily, but depends on the natural evolution of the temperature of the ceiling surface. The latter, in turn, is controlled by the mechanism of heat exchange between the ceiling and its surrounding environment. As discussed in Section 4.3.1, the temperature of the ceiling surface increases over time as more heat is transferred to it by means of convection and radiation. If initially the magnitude of heating greatly dominates the magnitude of cooling, the system ventilates upwards and becomes stratified into two layers, following the diagram in Fig. 4.22a. As the ceiling becomes warmer through various heat gains, convective heat exchange between the air in the room and the ceiling surface reduces, causing the relative magnitude of cooling to decrease. The temperature in the upper layer then increases, leading to an increase in positive buoyancy in the system. Consequently, the flow becomes faster and the upper layer depletes. This depletion of the upper layer causes thermal stratification to intensify over time. Eventually, when the temperature of the ceiling exceeds a certain value, the ceiling is rendered ineffective in providing comfort. Again, as in the case of the room heated by a distributed source, night cooling may be used to rejuvenate the system. If the night-time temperature of the exterior air is sufficiently low, the cooling provided by the newly rejuvenated ceiling may dominate the heating by the internal source at early occupancy hours on the next day. This may cause the room to become well-mixed and ventilate downwards initially, according to the diagram in Fig. 4.22c. Then, as the ceiling heats up during the day, its cooling flux may reduce sufficiently and become again dominated by the heating. As this occurs, the upward displacement flow resumes. Initially, the space remains well-mixed, as depicted in Fig. 4.22b because the cooling is still relatively strong. However, as the ceiling heats up sufficiently with time, the stratified temperature structure becomes re-established, taking the profile of Fig. 4.22a.

The mechanics of flows in a space cooled by a localised source at the ceiling and heated by a distributed source at the base may be readily understood by inverting the case in Fig. 4.22 geometrically.

4.3.3. Localised source of heat and localised source of cooling

Consider now the atrium in Fig. 4.24, which contains a concentrated group of people and a relatively small glass skylight through which heat is easily lost. The space is also equipped with low-level doorways and high-level openings. The basic natural ventilation flows that develop in this space may be understood by treating the glass roof as a localised source of cooling located on a flat ceiling and the occupants as a localised source of heat located on the floor. In addition, two openings, one at the top of the space and the other at the base, may

Figure 4.24. An atrium with a localised source of heat in the form of a concentrated group of people and a localised source of cooling in the form of a small glass skylight.

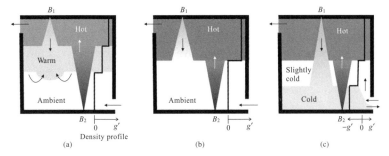

Figure 4.25. Steady flow regimes in a room with openings at two levels cooled by a localised source at the ceiling and heated by a localised source at the base, for when $B'_2/B_2 < 1$ (a), $B'_2/B_2 = 1$ (b) and $B'_2/B_2 > 1$ (c). (After Cooper & Linden 1996.)

be used to represent the windows and the doorways, respectively. For convenience, we may denote the buoyancy flux of the ceiling source by B_1, and that of the floor source by B_2. We may then describe the ratio of the two buoyancy fluxes by $\psi \equiv B_1/B_2 \leq 1$. Cooper and Linden (1996) showed that three ventilation regimes are possible in this situation, depending on the value of ψ.

If ψ is sufficiently small – that is to say, if the cooling is sufficiently weak compared with the heating – three-layered stratification, as sketched in Fig. 4.25a, develops. In this regime, the localised heat source at the base of the space produces a positively buoyant plume which rises in the space. This plume forms immediately underneath the ceiling a well-mixed layer of temperature above that of the exterior air. This hot layer drains to the exterior environment through the top opening. Meanwhile, the localised source of cooling located on the ceiling produces a cold plume. This cold plume descends through the aforementioned hot layer while entraining air from that layer. Thus the cold plume becomes less cold as it travels downwards and, consequently, forms underneath the hot layer an intermediate layer of temperature between the hot layer and the exterior environment. The bottommost zone near the base of the space is supplied by neutrally buoyant ambient air, drawn into the space from the exterior through the lower opening. At the interface, penetrative convection may develop that is driven by the cold plume and that entrains fluid from the ambient layer to the intermediate layer.

It has been mentioned in our discussion of the plume theory in Section 3.2.1 that the buoyancy flux of a plume reduces as it rises through a stratified environment. This is precisely what happens to the hot plume in Fig. 4.25a as it travels through the intermediate layer into the upper layer. In fact, it is this reduced buoyancy flux of the hot plume relative to its original value at the source, denoted by B'_2/B_2 say, that determines the temperature structure in the space. When ψ is sufficiently small that $B'_2/B_2 < 1$, the flow regime described in Fig. 4.25a is established. However, when ψ is increased sufficiently so that $B'_2/B_2 = 1$, the negative source becomes strong enough to supply the intermediate layer with air of the same temperature as that of the ambient air. The lower interface thus disappears, and the room assumes two-layered stratification shown in Fig. 4.25b. For higher values of ψ such that $B'_2/B_2 > 1$, the cooling becomes so strong that it forms a cold layer at the base of the space, causing three-layered stratification as shown in Fig. 4.25c to develop. In this regime, the top layer drains through the upper opening as before, but at the bottom opening there is an exchange flow that allows the bottommost layer to drain while admitting ambient air into the space. This ambient air rises as a plume to supply the middle layer of density slightly greater than the ambient environment.

The aforementioned different flow regimes have direct impacts on thermal comfort and indoor air quality. Imagine a room located in an uncomfortably warm environment heated by a strong localised source, such as a heavy machine. The room is occupied, but the heat produced by the occupants is small compared with the heat produced by the primary source, that is, the machine. Therefore, in the absence of cooling, this room will become stratified essentially into two layers. The occupants sitting in the lower zone will experience discomfort owing to the severity of the exterior air temperature. In an attempt to bring about thermal comfort, cooling may be supplied locally from the top of the space, for example, through a diffuser located on the ceiling. If the magnitude of this cooling is insufficiently strong, the occupants in the lower zone will not be able to enjoy cool air. This is because the descending cold plume will either form a warm intermediate layer above the heads of the occupants, as depicted in Fig. 4.25a, or fill the lower occupied zone with air of the same temperature as that outside, as depicted in Fig. 4.25b. To achieve thermal comfort, the magnitude of cooling will have to be so large that a cold layer is able to form in the lower occupied zone, as shown in Fig. 4.25c. If the magnitude of cooling is just right, this lower zone will be deep enough to cover the entire depth of the occupied zone and attain a comfortable temperature. However, if the magnitude of cooling is exceedingly large, discomfort will ensue. In this case, if the cold layer covers the whole depth of the occupied zone, the occupants will feel chilly from head to toe. If the cold layer is shallow, they will experience a cold feet sensation owing to an excessive vertical temperature gradient.

Detailed quantification of the aforementioned temperature structures is quite complex and beyond the scope of this book. Nonetheless, some broad remarks may be made in relation to it. We note that the approaches used for treating flows developed in a room with no cooling containing a localised source or localised sources of identical sign discussed in Sections 3.2.3–3.2.5 may be applied also to the present problem. Indeed, Cooper and Linden (1996), using these approaches, were able to show how, in the flow regime in Fig. 4.25a, an increase in the relative flux of cooling, ψ, strengthens the penetrative convection and lowers the lower interface. They also showed how the relative temperature of the two buoyant layers depends on the ratio ψ of the source strengths, the effective vent area and the relative height of the interfaces.

As for the regime in Fig. 4.25b, thanks to the system being stratified into two layers, the flow structure is comparable to that achieved in the room heated by a single localised source discussed in Section 3.2.3. In fact, the expression for the net volume flow for the regime in Fig. 4.25b takes the same form as Eq. (3.2.3-i) given for the room heated solely by a localised source. However, in identifying the height of the interface in the present case, the entrainment of the descending cold plume must also be taken into account. Cooper and Linden (1996) showed that this makes the position of the interface dependent on the ratio ψ of the strengths of the positive and negative sources.

As to the regime in Fig. 4.25c, the depths and temperatures of the topmost and bottommost layers may be identified using essentially the same approach as that used for treating a plume rising through a stratified environment discussed in Section 3.2.5. The procedure for the present case, however, is less straightforward, thanks to the opposing buoyancy of the sources and the exchange flow across the lower opening.

5 Some common flow complications arising from more complex geometries

We have hitherto considered flows in a wide range of spaces of simple geometry. Situations, however, may arise in which more complex spatial configurations lead to flow complications of a kind that cannot be adequately explained by the fundamentals covered previously; the richness of flow dynamics often increases with the complexity of space geometry. Although it is impossible to discuss all geometries, it is practical to address some of those that are more frequently encountered, which we do in this chapter. We begin with flows in spaces with vents at more than two levels. Horizontal vents such as ventilation chimneys that open upwards are considered as well as vertical vents such as windows and doorways – a minor distinction, perhaps, but it will be seen that the two types of opening can lead to very different flow behaviours. Then, we turn to flows developed in multiple connected spaces, beginning with multi-storey buildings and moving on to multiple spaces connected laterally through a doorway. Through our analysis, some unexpected flow phenomena are revealed. These include the possibility of multiple flow regimes within a single set of boundary conditions and instability associated with flows through joined spaces. It will be shown that in acquiring insights into the mechanics of flows in complex geometries, consideration of hydrostatic balance, Bernoulli's theorem and the conservation of mass, thermal energy and buoyancy flux are all key, as they were in examining the mechanics of flows in simple geometries.

5.1. Openings at more than two levels

5.1.1. Multiple stacks

All the natural ventilation problems examined thus far have unique flow solutions; that is, a given set of boundary conditions leads to only one flow regime, one temperature structure and one flow rate. However, this is not always the case, and there are situations in which a single set of boundary conditions allows multiple flow solutions. Such situations can be found, for example, in spaces with multiple ventilation stacks.

Fig. 5.1 shows a sketch of the School of Slavonic and East European Studies building in Bloomsbury, London, UK. This building is of interest to us because, like the BedZED in Fig. 1.1, the Contact Theatre in Fig. 1.2 and the Auden Theatre in Fig. 4.9, it is an example of an increasing number of buildings in which stacks, or chimneys, are used as a means of ventilation. The numbers of stacks in these buildings vary, of course, but they all share fundamental behaviours. These fundamental behaviours may be explained with reference to a simple space equipped with just two stacks and a low-level

Figure 5.1. The School of Slavonic and East European Studies in London, UK, and its ventilation stacks.

opening akin to a doorway, as shown in Fig. 5.2a. This space is heated by a uniform source at the base analogous to distributed occupants or an underfloor heater. Note that, as shown in this figure, the stacks may not, in reality, protrude from the top of the building or look like stacks at all, and may form parts of the building's roof structure or envelope. Indeed,

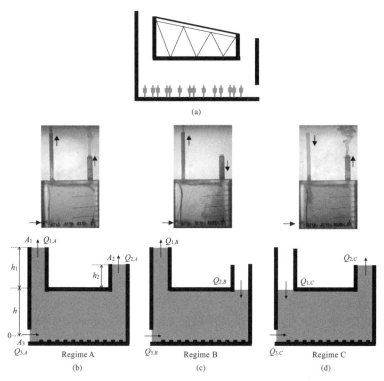

Figure 5.2. Space with two high-level ventilation stacks and a low-level opening heated by a distributed source at the base (a), and the three flow regimes that may develop in it at steady state (b–d). The shadowgraph images are taken from Chenvidyakarn and Woods (2005a).

Mingotti, Chenvidyakarn and Woods (2011) have shown that a double-skin facade that draws air from a low level and rejects it at a high level similar to that shown in Fig. 5.3 can also function as a stack.

Let us now consider Fig. 5.2a in more detail. Traditional intuition would suggest that this geometry could lead to only one steady-state flow regime, that is, one in which fresh air is

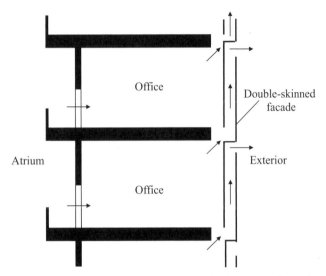

Figure 5.3. Double-skin facade as a ventilation stack. (Based on an idea by Mingotti, Chenvidyakarn & Woods 2011.)

drawn from the exterior into the space through the low-level opening while warm air is vented out through both stacks (Fig. 5.2b). This picture of flow – which is in essence the same as the simple displacement ventilation developed in a space heated uniformly but containing no stack discussed in Section 3.3.1 – is, however, incomplete. Water-bath experiments by Chenvidyakarn and Woods (2005a) revealed that, depending on the history of flow and the geometry of the space, up to three steady-state ventilation regimes are possible. In the first regime, the flow is all upward as described by Fig. 5.2b. (For convenience, let us call this regime A). In the second regime (called B say; Fig. 5.2c), the doorway acts as an inlet and the taller stack acts as an outlet as in the first regime, but the shorter stack admits inflow instead. In the third regime

(regime C; Fig. 5.2d), the doorway allows inflow as in the previous two regimes, but the shorter stack admits outflow and the taller stack admits inflow. In all the three regimes, the distributed heat source at the base of the space causes the interior to become well-mixed at steady state.

It is not difficult to imagine how the possibility of more than one flow regime can present a challenge for control. In fact, a number of control questions may arise. Will the flow rate and temperature vary from one regime to another? If so, how? When does a particular flow regime develop? Which regime is most conducive to comfort and energy efficiency? To answer these questions, analysis of the system needs to be carried out. This may be done based on a recognition that there is a neutral level that dictates the direction of flow through each opening (see Section 2.2). A complete mathematical model for the problem is given by Chenvidyakarn and Woods (2005a) and is not repeated here. It is sufficient for our present purposes just to note that an expression for the flow rate in each regime may be arrived at by tracing pressure along the streamline up to the level of each opening and considering the conservation of mass – the procedure not dissimilar to that presented in the introductory Section 2.5, in fact. The key difference between that introductory case and the problem in hand is that, with multiple stacks, the net buoyancy head driving the ventilation varies with the flow regime owing to the changes in the heights of the columns of hot air in the stacks as the direction of flow changes (compare Figs. 5.2b–d). This, in turn, leads to variation in the definition of the net volume flux as the flow regime changes. For example, in regime A, the total volume flux is identical to the inflow volume flux from the low-level opening, which in turn is equal to the sum of outflows through both stacks, whereas in regime B, the net volume flux is equal to the outflow volume flux through the taller stack, which itself is equal to the sum of inflows through the lower opening and the shorter stack. With this complication recognised, the steady-state interior temperature may then be found by considering the balance between heat loss driven by the net volume flux in each regime and heat input from the source at the base of the space.

Owing to the variation in the definition of the net volume flux and the net buoyancy head driving the ventilation flow, for a fixed geometry and heat flux, each flow regime usually leads to a different net flow rate and a different interior temperature (Fig. 5.4). In general, regime A, with the flow all upward, results in the fastest net flow and the coolest room, except when the size of the lower opening relative to those of the other openings is small. Of regimes B and C, the former, venting through the taller stack, thereby having a greater buoyancy head, produces a faster flow and a cooler room. To illustrate these statements with some figures: a building may be heated by a source of about 8000 W in output (equivalent to about 80 occupants), have a floor-to-ceiling height of 10 m, and be connected to two stacks of heights 10 and 7.5 m, each with a cross-sectional area of about 2 m². Depending on the size of its lower opening, this building may have almost 20% more flow and about 5% lower temperature when ventilated in regime A than when ventilated in regime C. It may have about 10%

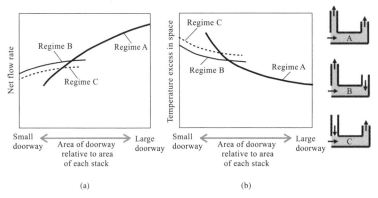

Figure 5.4. Variation in the net flow rate (a) and the temperature excess in the space (b) across the three flow regimes as a function of the geometry of the space. In plotting these curves, the sizes of both stacks and the heat flux are taken as identical in all regimes, and a certain ratio of stack heights is assumed. The overlapping of two or more curves indicates the possibility of multiple flow regimes. (Based on model and experimental data by Chenvidyakarn & Woods 2005a.)

more flow and a few percent lower temperature when ventilated in regime B than when ventilated in regime C.

These variations in the flow rate and interior temperature may seem trivial, but the human body is a sensitive comfort measuring device: a change by one or two degrees in temperature or a small drop in the ventilation rate could make a difference between a pleasant environment and one hard to bear. That multiple flow regimes are possible does not necessarily mean that all the flow regimes lead to thermal comfort and sufficient ventilation. In fact, in most cases, only a certain flow regime or regimes and a certain range of vent sizes and heights allow these ventilation goals to be achieved. The possibility of multiple flow regimes, therefore, must be taken into consideration when developing a control strategy for the ventilation system.

In practice, the choice of the flow regime tends to depend strongly on how many people and how much heat-emitting equipment there are in the building. This is because the absolute values of the flow rate and the interior temperature achieved in each regime are influenced to a large degree by the size of the heat load. A flow regime that delivers thermal comfort and sufficient ventilation at one size of heat load may fail to do so at another size. This is illustrated by the plot in Fig. 5.5. It can be seen that, broadly speaking, regime A, providing the fastest flow and therefore the greatest rate of heat removal, is required when the heat load is large, whereas regimes B and C are also viable at smaller heat loads. However, it would be wrong to conclude from this that regime A is the surest one to deliver comfort, and that a building should be designed solely or primarily to achieve this regime. In a situation where the temperature of the exterior air is sufficiently low, fast ventilation effected by regime A would lead to an uncomfortably low interior temperature and excessive energy loss, whereas regime B or C, with its lower rate of flow and higher temperature, could provide satisfactory interior conditions.

It should be obvious by now that knowing when each regime develops is key to developing effective controls for the system. Figs. 5.4 and 5.5 have already alluded to the fact that geometrical relations play a part in controlling the ventilation regime; it will be useful to add some precision here. Chenvidyakarn and Woods (2005a) have shown theoretically

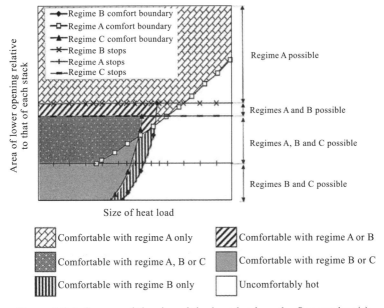

Figure 5.5. Influence of the size of the heat load on the flow regime(s) required to achieve thermal comfort and sufficient ventilation. (Based on calculations by Chenvidyakarn & Woods 2005a.)

and experimentally that the all-upward displacement flow, that is, regime A, is possible only when the lower opening is sufficiently large for the height of the taller stack that the inflow through the lower opening alone is adequate to supply outflow through both stacks. For a system with same-sized stacks, this condition is satisfied when

$$\frac{A_3}{A_1} \geq \frac{(h_1 - h_2)^{1/2}}{(h + h_2)^{1/2}}, \qquad (5.1.1\text{-i})$$

where $A_1 = A_2$ is the effective area of each stack, and A_3 is the effective area of the lower opening. Each effective opening area is defined according to Eq. (2.4-i). The variable h is the height of the ventilated space as measured from the midpoint of the lower opening to the bases of the stacks, h_1 is the height of the taller stack, and h_2 is the height of the shorter stack. Equation (5.1.1-i) shows that if the stacks are symmetrical, $h_1 = h_2$, regime A is possible at all sizes of the lower opening. On the contrary, if the stacks are asymmetric, that is, $h_1 \neq h_2$, and if the lower opening is exceedingly small, one stack develops inflow to complement small inflow from the lower opening, leading to either regime B or C, depending on the relative height of the stacks and the size of the lower opening relative to those of the stacks. It may be shown theoretically and experimentally that regime B is possible when

$$\frac{A_3}{A_1} \leq \left(\frac{h_1}{h}\right)^{1/2}, \qquad (5.1.1\text{-ii})$$

and that regime C is possible when

$$\frac{A_3}{A_1} \leq \left(\frac{h_2}{h}\right)^{1/2} \frac{A_2}{A_1}. \qquad (5.1.1\text{-iii})$$

One thing is immediately clear from Eqs. (5.1.1-i)–(5.1.1-iii): the ventilation regime is a function of the geometry of the space alone and is entirely independent of the size of the heat load. The geometrical limits of the three flow regimes as given by Eqs. (5.1.1-i)–(5.1.1-iii) are plotted in Fig. 5.6. It may be seen that for certain ranges of stack heights and vent sizes, a unique ventilation regime is obtained, but for others, two or even three ventilation regimes are possible. Numerical values will give some idea as to the sizes of vents required to achieve a desired ventilation regime in actuality. An office lobby may

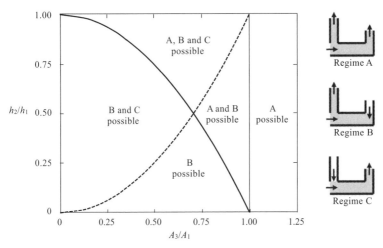

Figure 5.6. Limits of the three ventilation regimes as a function of the relative size of the openings, $A_3/A_1 = A_3/A_2$, and the relative height of the stacks, h_2/h_1. (Based on model by Chenvidyakarn & Woods 2005a.)

have a floor-to-ceiling height of 4 m and may be connected to two stacks of heights 4 and 3 m. The ratio of the heights of these stacks is therefore 0.75. If the areas of the openings of the two stacks are equal at say 1 m², then to maintain all-upward ventilation (regime A), its lower vent (which may take the form of a doorway) must be opened to make up an effective area of at least 1 m². On the other hand, if regime C is to be achieved, the lower vent will have to be no greater than about 0.85 m².

The question now is how we can ensure the desired ventilation regime when a number of regimes are possible (e.g., when the relative size of the lower opening of the building in Fig. 5.6 is less than unity). Water-bath experiments by Chenvidyakarn and Woods (2005a) suggested that this may be

accomplished by means of appropriate interventions. Closing the inflow stack in regime B or C and allowing it to heat
up sufficiently before opening it again can bring about regime
A, whereas using a fan to push air in one of the stacks down in
regime A can trigger the development of regime B or C. These
interventions, it may be noted, need only be temporary – that
is, they may be removed once the desired ventilation regime
is established – thanks to the system being capable of achieving multiple steady-state flow regimes within a single set of
boundary conditions. This is different from a system that has
a unique flow solution, in which a permanent change in the
geometry of the space is required to change the flow regime
at steady state.

Finally, it is worth pointing out that even when the distributed heat source at the base of the space is replaced by a
localised source, the possibility of multiple flow regimes will
persist, although the picture of flow will be different from that
described above. Results from Cooper and Linden (1996) and
Linden, Lane-Serff and Smeed (1990) (discussed in Sections
4.3.3 and 3.2.3, respectively) suggest that the presence of a localised source will lead to the space becoming stratified instead
of well-mixed. In these conditions, if the ventilation regime
that develops is equivalent to regime A – that is, the flow is all
upward – the hot plume from the localised source will form
a buoyant layer immediately underneath the ceiling, leaving
the lower zone filled with ambient air (Fig. 5.7a) – in other
words, the same flow configuration as the classical plume ventilation described by Linden, Lane-Serff and Smeed (1990)
will develop. In contrast, if the room attains an equivalent of

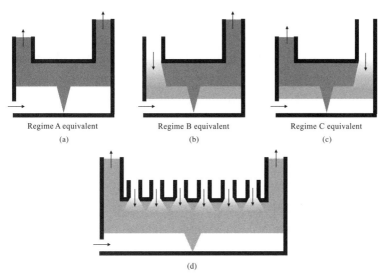

Regime A equivalent
(a)

Regime B equivalent
(b)

Regime C equivalent
(c)

(d)

Figure 5.7. Steady-state flow regimes which may develop in spaces of different stack configurations heated by a localised source.

regime B or C, there will be inflow through one or more of the stacks. Fresh air entering the space through these stacks will descend through the upper layer as plumes, entraining warm air from that layer along the way. These plumes will become progressively buoyant and, on reaching the bottom of the upper layer, will form an intermediate layer of temperature between the upper layer and the ambient air. In this way, three-layered stratification will develop instead of two (Figs. 5.7b and c). The depth and temperature of this intermediate layer will depend chiefly on the entrainment of the descending plume. If the entrainment is small, the layer will be thin and attain a temperature close to that of the ambient environment. If the entrainment is strong, the layer will be deep and attain a temperature close to that of the upper layer. However, when there are a large number of stacks well distributed across the space, the descending plumes may be able to mix fully with the air in the upper layer. If this happens, an intermediate layer will not be formed, and the room will become stratified into just two layers (Fig. 5.7d). In any of these cases where the space is stratified, the flow regime may or may not have an impact on thermal comfort, depending on the position of the interface between the lowest buoyant layer and the ambient layer underneath. If the lowest interface lies within the occupied zone, variation in the flow regime, which affects the temperature in the buoyant layer(s) but not the temperature in the ambient zone below, will have an influence on the level of comfort perceived by the occupants. However, if the lowest interface lies above the occupied zone so that the occupants experience only cold air from the exterior, the flow regime will have no influence on the perceived indoor air temperature whatsoever. In this case, any change in the flow regime will affect only the flow rate and indoor air quality.

5.1.2. Multiple side openings

In certain situations where the inclusion of high-level stacks would interfere with the overall aesthetics of the building or incur excessive costs, natural ventilation could be accomplished instead through side openings located at high levels up the facade of the building. Modern buildings with such features abound: an atrium may have windows open at multiple levels several storeys above the ground, or a room may be

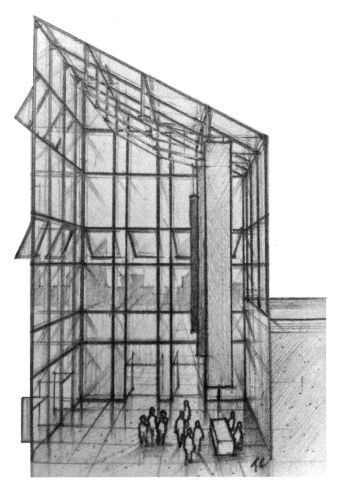

Figure 5.8. An example of a modern space containing side openings at more than two levels.

fitted with a combination of a low-level doorway, a mid-level window and a high-level clerestory opening (e.g., Fig. 5.8). It is shown in this section that just by having vents open sideways instead of upwards as in the case of stacks, the behaviour of

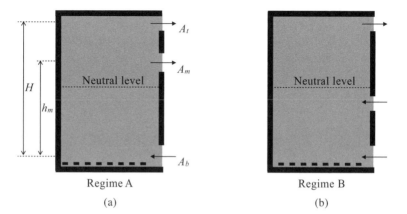

Figure 5.9. Two steady-state flow regimes that can develop in a space with side vents at three levels, when the space is heated by a uniform source at the base. (Following Fitzgerald & Woods 2004.)

flow in the space becomes fundamentally different, and that a different control strategy is required.

Fig. 5.9 shows a schematic of a space fitted with vertical openings at three different levels. Again, we use a building with vents at three levels as the basis for our examination because it is the simplest geometry that can shed light on the fundamentals of flow associated with vents at n levels. Fitzgerald and Woods (2004) showed that if this space is heated by a distributed source at the base, it becomes well-mixed at a temperature above that of the exterior air at steady state. In these conditions, the temperature excess in the space produces pressure gradients across the three openings which dictate the direction of flow through each opening (cf. Section 2.2). At the base of the space, pressure deficiency in the interior draws inflow through the bottom opening. At the top of the space, pressure excess in the interior drives outflow through the top opening. At some height above the base, a neutral level is established at which pressure between the interior and exterior are equal. The position of this neutral level relative to the position of the middle vent dictates the direction of flow through the middle vent. If the neutral level lies below the middle vent, there will be outflow (Fig. 5.9a), whereas if the neutral level lies above it, there will be inflow (Fig. 5.9b). For convenience, let us call the former flow configuration regime A, and the latter regime B.

Do we have again, then, a system capable of producing multiple flow regimes within a single set of boundary conditions as in the case of stack ventilation discussed earlier? An examination of the limits of the flow regimes will quickly answer this question. Figure 5.9 clearly shows that the flow regime changes from one to the other when the height of the neutral level becomes identical to that of the middle vent. Fitzgerald and Woods (2004) shows that this condition is met when

$$\left(\frac{h_m}{H}\right)^{1/2} = \frac{A_t}{A_b}\left(1 - \frac{h_m}{H}\right)^{1/2}, \qquad (5.1.2\text{-i})$$

where h_m is the height of the middle vent as measured from the midpoint of the lower vent, H is the vertical separation between the bottom and top vents, A_t is the effective area of the top vent taking into account pressure loss at the vent, and A_b is defined similarly for the bottom vent. The limits of the flow regimes as given by Eq. (5.1.2-i) are plotted in Fig. 5.10. It can be seen clearly that, for a given set of boundary conditions, the flow solution is *unique* – that is, a single set of boundary conditions leads to only one flow regime. This is in contrast to the case of stack ventilation, in which multiple flow solutions are possible. However, as with the case of stack ventilation, the flow regime accomplished through side vents is controlled by the geometry of the space and not the magnitude of the internal heat flux.

An immediate implication of these results is that achieving the desired ventilation regime in a space with multiple side vents is generally easier than in a space with multiple stacks: so long as the vents are positioned and sized appropriately, the desired ventilation regime will always be achieved, without the need to coerce the flow through temporary interventions. For example, Fig. 5.10 shows that for a system with equal top and bottom vent sizes, $A_t/A_b = 1$, the neutral level is halfway up the height of the space, $h_n/H = 0.5$. Therefore, to achieve regime A, one simply positions the middle vent above this level, whereas to achieve regime B, one simply places the middle vent below it. This critical height of the middle vent that is the boundary of the flow regimes rises as the ratio A_t/A_b of the sizes of the top and bottom vents increases. This is because as the top vent gets relatively larger, a greater proportion of pressure loss through the system occurs across the bottom vent so that the neutral level moves upwards. This shift in the neutral level can be quite sensitive to a change in the opening ratio. It may rise by almost 40% when the relative size of the top vent is increased by 50% from unity. Note that the size of the middle vent does not enter into the determination of the flow regime, because at the point of regime change the flow through this vent is zero.

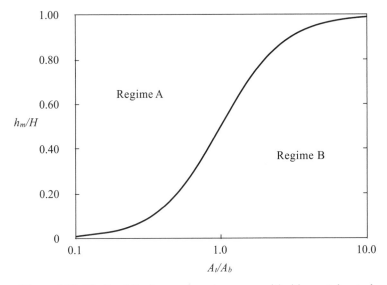

Figure 5.10. Limits of the flow regimes in a space with side vents located at three levels heated by a uniform source at the base. The curve indicates the neutral buoyancy position, h_n/H. (Based on model by Fitzgerald & Woods 2004.)

A question may now arise as to which of the two regimes is more conducive to thermal comfort and good ventilation. The answer is that, at the *building scale*, neither regime is inherently better than the other, provided that the geometries and the heat fluxes in the two regimes are comparable. For example, when the sizes and heights of the top and bottom vents are identical, the neutral position will be half-way up the height of the space in both regimes. In these conditions, if the middle vent is located the same distance away from the neutral level in regime A as it is in regime B, there will be the same amount of pressure driving air through the three vents in both regimes. Therefore, for the same heat flux, the flow rate and the interior temperature in both regimes will be identical.

However, it would be wrong to say that there are no circumstances at all in which one regime performs better than the other. Although it is true that at the *building scale* the two regimes affect thermal comfort and ventilation identically, at a *local scale* their effects could be different. For example, a source of airborne contaminants may be located at some height above the base of the space next to the middle vent. In this case, ventilating the space in regime A could help remove the contaminants before they spread substantially indoors, whereas ventilating the space in regime B could encourage their dispersion into the occupied zone, adversely affecting indoor air quality.

To control the flow rate and the interior temperature in the two regimes the same strategy may be used. Obviously, increasing the area of any opening in any of the regimes will lead to an increase in the flow rate and a reduction in the interior temperature. Less obvious perhaps is that the flow rate and the interior temperature can also be controlled via the positioning of the middle vent: moving the middle vent upwards or downwards away from the neutral level will increase the flow rate and reduce the temperature. This is because doing so will increase the hydrostatic pressure driving the flow through the middle vent, which in turn increases the rate of heat removal from the space. Depending on the circumstances, the effects of the movement of the middle vent can be quite considerable. At larger sizes of the middle vent, a shift by 40% or so in its position away from the neutral

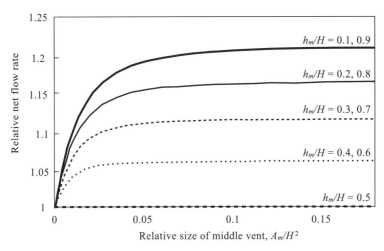

Figure 5.11. Effects of the position and the size of the middle vent on the flow rate. In this plot, the top and bottom vents are assumed to be of the same size so that the neutral level is halfway up the height of the space, $h_n/H = 0.5$. Regime A develops when the middle vent lies above this level, $h_m/H > 0.5$, while Regime B develops when the middle vent lies below this level, $h_m/H < 0.5$. The flow rate is shown as relative to the flow rate achieved when the middle vent is at mid-height, $h_m/H = 0.5$. (Based on calculations by Fitzgerald & Woods 2004.)

level may lead to an increase in the ventilation rate by just over 10% and a reduction in the interior temperature by a similar amount (Fig. 5.11). This is equivalent to a decrease in temperature of over 2°C in a room originally at 25°C, a change substantial enough to make a difference between a comfortable environment and an uncomfortable one.

At this point, a question may be asked as to what will happen if the room is heated by a localised source instead of a distributed source. The work by Linden, Lane-Serff and Smeed (1990), built on the plume theory of Morton, Taylor

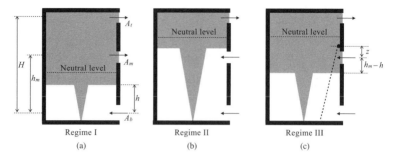

Figure 5.12. Steady-state flow regimes which develop in a space with side vents located at three levels, when the space is heated by a localised source at the base. (Following Fitzgerald & Woods 2004.)

and Turner (1956), provides a basis for answering this question. As with the case of a room with vents at two levels heated similarly discussed in Section 3.2.3, the room with vents at three levels generally becomes stratified into two layers (a departure from this is discussed later in this section). The localised plume forms a warm buoyant layer in the upper part of the space while cold ambient air fills the lower zone. However, now there is also flow through the middle vent. The direction of this flow depends on where the neutral level is relative to the height of the middle vent. If the neutral level lies below the middle vent, there is outflow through the vent. The system in this case operates in regime I shown in Fig. 5.12a. In contrast, if the neutral level lies above the middle vent, there is inflow through the vent. In this case, two flow regimes are possible, depending on the position of the interface between the warm and ambient layers relative to the position of the middle vent. If the interface lies above the middle vent, the inflow from the middle vent goes into the ambient zone and regime II in Fig. 5.12b develops. If the interface lies below the middle vent, the inflow from the middle vent goes into the buoyant layer and regime III in Fig. 5.12c develops.

So how do these different flow regimes affect thermal comfort and ventilation quality? An immediate conclusion that may be drawn from the diagrams in Fig. 5.12 is that, as the flow regime changes, the flow rate also changes. This is evident from the movement of the interface. We recall from the plume theory in Section 3.2.1 that a larger flow leads to a higher interface. Therefore, regime II, with the highest interface, produces the fastest flow, providing the most effective removal of airborne contaminants, followed by regime III and regime I. For a fixed heat flux, the faster the flow is, the lower the temperature in the buoyant layer is.

However, this does not mean that regime II is always the best of the three. In certain circumstances, for example, in winter, the flow rate may need to be minimised to avoid excessive heat loss. In these cases, regimes I and III may be more conducive to comfort and energy efficiency. This is not only because the flow rates achieved in these regimes are smaller, but also because the lower positions of the interface allow the occupants, who usually occupy a zone near the base of the space, to enjoy more of the warm air of the upper layer.

Clearly, to achieve the required flow regime, it is important to know the condition for each regime to develop. It can be seen directly from Fig. 5.12 that the transition from regime III to regime I takes place when the neutral level descends to the level of the middle vent. It may be deduced from the work by Fitzgerald and Woods (2004) that this occurs when

$$\frac{h_m}{H} = \frac{\frac{h}{H} + \left(\frac{A_t}{A_b}\right)^2}{1 + \left(\frac{A_t}{A_b}\right)^2}, \tag{5.1.2-ii}$$

where h is the height of the interface given by the relation

$$\frac{h}{H} = \frac{\left(1 - \frac{h_m}{H}\right)^{1/5} \left[2\left(\frac{A_t}{H^2}\right)^2\right]^{1/5}}{\lambda^{3/5}}. \tag{5.1.2-iii}$$

The variables A_t and A_b are the area of the top and bottom vent taking into account pressure loss at each vent respectively, h_m is the height of the middle vent, H is the height of the entire system, and λ is the integrated constant describing the entrainment of the plume given by Eq. (3.2.1-xiii).

Furthermore, regime II changes to regime III when the interface between the warm and ambient layers descends to the level of the middle vent. Drawing on the work by Fitzgerald and Woods (2004), it can be shown that this happens when

$$\frac{\left(\frac{h_m}{H}\right)^5}{\left(1 - \frac{h_m}{H}\right)} = \frac{1}{\lambda^3}\left(\frac{A^{**}}{H^2}\right)^2, \tag{5.1.2-iv}$$

where A^{**} is the effective opening area given by

$$A^{**} = \frac{A_t\left(\frac{A_m}{A_b} + 1\right)}{\left[\frac{1}{2}\left(\frac{A_t^2}{A_b^2} + \left(\frac{A_m}{A_b} + 1\right)^2\right)\right]^{1/2}}, \tag{5.1.2-v}$$

with A_m being the area of the middle vent which takes into account pressure loss at the vent. The expression (5.1.2-iv) indicates that, unlike when the room is heated uniformly, when the room is heated locally the effective vent size also enters into the determination of the flow regime (cf. Eq. (5.1.2-i)). This is because the effective vent size controls the flow rate, which in turn controls the interface height. The neutral level, being always above the interface, may rise or fall above or below the middle vent following the movement of the interface. Note, however, that as with the case of distributed heating, the flow regime in the case of localised heating is controlled by the geometry of the space but not the heat flux (Eqs. (5.1.2-ii) and (5.1.2-iv)).

At this stage, it is important to point out that the diagram of regime III in Fig. 5.12c represents but one possible flow structure that can develop when there is inflow to the upper layer. This diagram assumes that when the plume enters the upper layer and descends in it, it barely mixes with the air therein. Therefore, all the inflowing fresh air effectively descends to the base of the space, leading to simple two-layered stratification. However, this may not always be the case. Depending on its strength, the descending plume may entrain to a noticeable degree the air in the upper layer. If this happens, it will become progressively warmer as it descends, and on reaching the interface between the upper and lower layers, it will attain a temperature between the two layers. As a result, its descent will be arrested, and it will spread underneath the upper layer, causing three-layered stratification to develop, as depicted in Fig. 5.13a (case III.x). However, if the mixing between the descending plume and the air in the upper layer is very vigorous, the descending plume may be able to

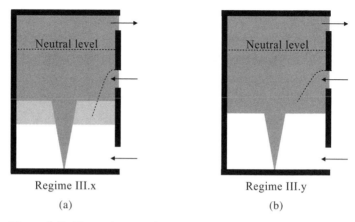

Regime III.x   Regime III.y

(a)   (b)

Figure 5.13. Two other steady-state flow regimes which may develop when there is inflow from the middle vent to the upper layer. In (a), the inflowing air partially mixes with the air in the upper layer to form an intermediate layer. In (b) the inflowing air mixes completely into the upper layer so that no intermediate layer is formed. (Following Fitzgerald & Woods 2004.)

mix completely with the air in the upper layer. If this happens, no fresh air will descend past the interface between the upper and lower layers, and two-layered stratification will develop instead, as shown in Fig. 5.13b (case III.y).

These three different flow structures can have direct impacts on indoor thermal comfort. When the plume barely mixes with the upper layer (regime III), heat supplied by the localised source is confined within the upper zone. If this upper zone lies above the occupied level, the occupants will not benefit from the heating. However, if the plume entrainment is sufficiently strong that an intermediate layer develops (regime III.x), the occupants may benefit from the heating through the intermediate layer, which carries some heat from the upper layer down to a lower level, provided that the inter-mediate layer is established low enough to be within the occu-pied zone. On the other hand, when the entrainment is suffi-ciently vigorous that the descending plume mixes completely with the air in the upper layer (regime III.y), the upper layer may become quite cold. In this situation, even if the upper layer lies within the occupied zone, the occupants may feel discomfort.

Quantification of flows in the three regimes requires dif-ferent approaches. In regime III.y, this is quite straightfor-ward and may be carried out based on the simple conserva-tion of mass and thermal energy, as shown by Chen and Li (2002). For regime III.x, the entrainment of the descending plume may be estimated using the approach due to Cooper and Linden (1996) discussed towards the end of Section 4.3.3. As to regime III, its limit may be defined according to Woods, Caulfield and Phillips (2003) as follows. The air entering the middle vent may be treated as having a finite mass flux and originating from a virtual source located a distance z above the middle vent. Regime III develops if the amount of air in the upper layer that would be entrained by the plume as it descended over the distance z were much greater than the amount of air in the upper layer that is actually entrained as the plume descends over the distance between the middle vent and the interface – in other words, regime III only develops when the entrainment of the plume is minimal. Because the amount of air entrained scales with plume height (Morton, Taylor & Turner 1956; Eq. (3.2.1-v)), we may infer that the above condition is satisfied when $(h_m - h)/z \ll 1$, or when the interface lies very close to the middle vent (with h_m being the height of the middle vent and h the height of the interface, both measured from the lower opening). The virtual origin height z may be found by equating the theoretical volume flux of the descending plume at the level of the middle vent with the actual volume flux entering the same vent.

5.2. Multiple connected spaces

5.2.1. Multi-storey buildings

An increasingly common sight in the present-day environment is multi-storey naturally ventilated modern buildings. In these buildings, it is customary to see spaces on different floors interconnected via shared spaces, such as atria and open wells. Examples of these buildings include the International Digital Laboratory in Coventry, UK, shown previously in Fig. 3.24 and the Heelis office of the National Trust in Swindon, also in the UK, shown in Fig. 5.14. In the former, a first floor concourse is linked to open-plan offices on higher and lower floors and a ground floor lobby by a series of open wells. In the latter, open-plan offices on higher floors are connected to a lobby and a cafeteria on the ground floor via a central atrium. In each of these buildings, each floor contains its own heat loads and ventilation openings in the form of windows, doorways and/or openable skylights. Natural ventilation flows that develop in these two buildings, as with those developed in many other buildings like them, are quite complex, with the level of complexity tending to increase with the numbers

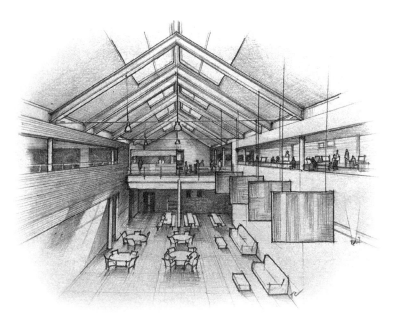

Figure 5.14. The National Trust's Heelis building in Swindon, UK.

of storeys, heat sources and openings involved. Nonetheless, these flows share basic principles. These principles may be understood by examining a relatively simple building containing just two storeys linked via an atrium located on one side, with each storey heated by a distributed source analogous to distributed occupants in an open-plan space. This is the scenario described by the solid outline in Fig. 5.15. The dotted outline in the same figure gives an idea of the possible richer geometry to which the basic flow principles obtained from the analysis of simpler geometry also apply.

An entry to the problem of flow in multi-storey buildings may be gained by revisiting the ventilation of a single space heated by a uniform source at the base examined in

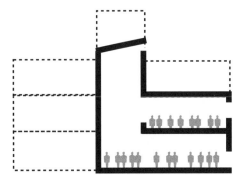

Figure 5.15. Building with multiple storeys linked via a common atrium.

Section 3.3.1. This is equivalent to assuming that the building in Fig. 5.15 has windows at just two levels, on the lower floor and at the top of the atrium, with the heat load located on the lower floor only (Fig. 5.16a). The results from the work by Gladstone and Woods (2001) discussed in Section 3.3.1 immediately allow us to deduce that the heat load

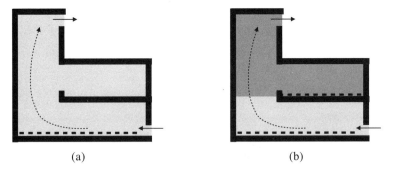

        (a)   (b)

Figure 5.16. Displacement ventilation regimes arising in a multi-storey building, when there are openings at two levels and a heat load on the lower floor only (a), and when there are openings at two levels and heat loads on both floors (b). (Following Gladstone & Woods 2001 and Livermore & Woods 2007.)

on the lower floor will lead to the whole building becoming well-mixed at a temperature above that of the exterior air at steady state, and that, at this state, there will be displacement ventilation drawing inflow through the lower floor window while driving outflow through the atrium opening.

Now if we extend the problem by adding a uniform heat load to the upper floor, the picture of flow becomes different (Fig. 5.16b). Even though upward displacement ventilation is maintained, the building becomes stratified into two layers rather than well-mixed at steady state (Livermore & Woods 2007). The lower zone, from the level of the lower floor to the level of the upper floor, attains a uniform temperature above that of the exterior air. The upper zone, from the level of the upper floor to the top of the atrium, attains a uniform temperature above that of the lower zone. This stratification is a result of successive heating as fresh air enters the building and rises up the atrium to the upper part of the building.

The problem may be extended further by adding a window to the upper floor (Fig. 5.17). Livermore and Woods (2007) showed that the flow structure in this case becomes much less straightforward than before, and several flow regimes are now possible (Figs. 5.17a–c). If the neutral level lies between the lower and upper floor windows, the upper floor window allows outflow. Ventilation mode A then develops, as shown in Fig. 5.17a. However, if the neutral level lies between the upper floor window and the atrium opening, the upper floor window acts instead as an inlet. Now two ventilation modes are possible, depending primarily on the relative strength of the heat loads on the two floors. If both heat loads

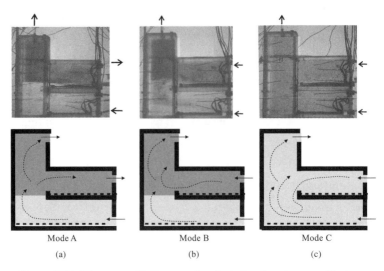

Figure 5.17. Three ventilation modes that develop in a multi-storey building whose floors are connected via an atrium and ventilated through independent openings. The shadowgraphs are taken from Livermore and Woods (2007).

are of comparable magnitude, the heat load on the upper floor is capable of offsetting the cooling by the inflow of ambient air through the upper floor window. As a result, the system becomes stratified into two layers, and ventilation mode B in Fig. 5.17b develops. However, if the heat load on the upper floor is significantly smaller than that on the lower floor, it may not be sufficient to heat the cold inflow from the upper floor window to a temperature greater than that in the lower zone. Consequently, cold air from the upper zone sinks to the lower zone. The two fluids are then mixed by the distributed heat source on the lower floor to a uniform temperature. Therefore, the entire building becomes well-mixed at a temperature above that of the exterior air at steady state and assumes ventilation mode C shown in Fig. 5.17c.

The transitions between the three modes of ventilation may be understood by considering the diagrams in Fig. 5.17. It can be seen clearly that mode A ceases to exist and mode B begins when the position of neutral buoyancy rises to the level of the upper floor window, so that the outflow through the upper floor window becomes inflow. Livermore and Woods (2007) showed that this transition occurs when

$$\gamma > \left[\frac{H^* \left(h^* - h_2^*\right) - h_2^*}{\left(H^* + 1\right)\left(h_2^* - 1\right)} \right]^{\frac{1}{2}}, \qquad (5.2.1\text{-i})$$

where γ is the ratio of the effective area of the atrium opening to the effective area of the lower floor window, H^* is the ratio of the upper floor heat load to the lower floor heat load, h^* is the vertical position of the upper floor heat load relative to the vertical separation between the lower floor window and the atrium opening, and h_2^* is the vertical position of the upper floor window relative to the vertical separation between the lower floor window and the atrium opening.

Further, mode B changes into mode C when the temperature in the upper zone becomes equal to that in the lower zone so that internal stratification ceases to hold. Livermore and Woods (2007) showed that this condition is met when

$$\gamma > \left[\frac{\lambda^2 \left(1 + H^*\right)^2 h_2^*}{\lambda^2 \left(1 - h_2^*\right) - H^{*2}} \right]^{\frac{1}{2}}, \qquad (5.2.1\text{-ii})$$

where λ is the ratio of the effective area of the upper floor window to the effective area of the lower floor window.

Some important points on control may be immediately drawn from inspecting Eqs. (5.2.1-i) and (5.2.1-ii). First, the size of the upper floor window does not enter into the determination of the transition from mode A to mode B. This is because at the point of transition between these two modes, the flow through the upper floor window is zero. However, the size of the upper floor window does affect the transition between mode B and mode C. This is because this window influences the volume flow through the upper zone and the heat balance therein. Furthermore, with the system containing two buoyancy sources, regime change becomes dependent on the relative strength of the sources, as well as on the geometry of the space. This result is to be compared with those in the cases of stack ventilation and multiple side openings discussed in Section 5.1, where only one buoyancy source is present and regime change is independent of the strength of buoyancy.

The limits of the flow regimes as given by Eqs. (5.2.1-i) and (5.2.1-ii) are plotted in Fig. 5.18. In this plot, the heat load height of $h^* = 0.5$ and the upper floor window height of $h_2^* = 0.6$ are taken; however, the general principles captured by the plot also apply to other cases with different values of h^* and h_2^*. What we can directly appreciate from this plot is that, in a multi-storey space, each set of boundary conditions admits a unique flow solution. This implies that to effect a change in the flow regime, a change in the geometry of the space or the heating conditions is required. Furthermore, it may be seen that mode A becomes more difficult to achieve as the relative size of the atrium opening increases (so that γ increases).

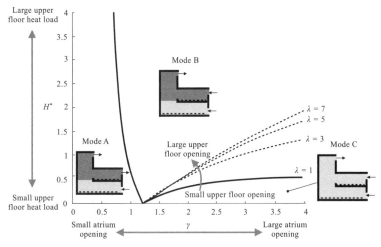

Figure 5.18. Limits of ventilation regimes. (Based on calculations by Livermore & Woods 2007.)

This is because as the atrium opening gets larger, outflow through the top of the building is enhanced, which in turn requires larger inflow to match. If the sizes of the upper and lower floor windows are fixed, this makes the direction of flow through the upper floor window likely to be inflow rather than outflow, making for conditions conducive to mode B. Mode B, however, becomes more difficult to maintain as the size of the upper floor opening increases (so that λ increases). This is because an increase in the size of the upper floor window enhances inflow through the upper zone. This reduces the temperature in the upper zone towards that in the lower zone, making for conditions conducive to mode C.

So how do these different modes of ventilation affect thermal comfort and indoor air quality? Is there any particular mode that is better than the others? In most circumstances, mode B is probably the most conducive to good indoor air quality, because in this mode the occupants on each floor receive fresh air from the exterior while stale air from each floor is vented out through the atrium. In contrast, in mode A, the occupants on the upper floor receive foul air from the lower floor, which could lead to poor indoor air quality. Likewise, in mode C, the occupants receive foul air from the surrounding floors, thanks to the well-mixed interior.

In terms of temperature and flow rate, mode A produces the slowest ventilation and the strongest stratification for a fixed combination of heat fluxes (Fig. 5.19). This is due to the fact that quite a small atrium opening is required to achieve this mode of ventilation (see Fig. 5.18), which makes for a small effective vent size if the sizes of all other openings are held constant across all modes. Mode B, on the other hand, requiring the atrium opening to be sufficiently large to develop (Fig. 5.18), produces a faster flow and weaker stratification (Fig. 5.19). However, as with mode A, mode B is prone to overheating on the upper floor, especially when the exterior air temperature is high or when the upper floor is heavily occupied, thanks to the stratified thermal structure. Increasing the relative size of the atrium opening can help alleviate this problem because doing so enhances the inflow through the upper floor window (Fig. 5.19a) so that, if the heat loads in the building are fixed, the temperature on the upper floor reduces (Fig. 5.19b). However, as a side effect of this, the inflow through the lower floor window reduces, due to the mass balance of the system (Fig. 5.19a). This causes

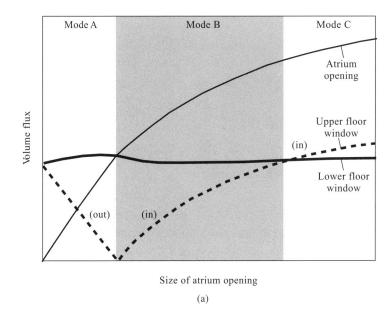

Size of atrium opening

(a)

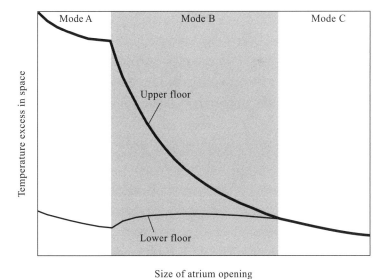

Size of atrium opening

(b)

Figure 5.19. Effects of the ventilation mode on the flow rate (a) and the interior temperatures (b). In these plots, the sizes of the upper floor and lower floor windows and the heat loads are held constant. (Based on calculations by Livermore & Woods 2007.)

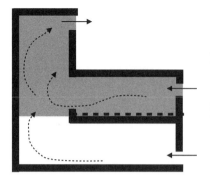

Figure 5.20. Ventilation of the lower floor using buoyancy borrowed from the upper floor. (Following Livermore & Woods 2006.)

the temperature on the lower floor to increase, approaching that on the upper floor (Fig. 5.19b). Therefore, care must be exercised in a few respects when enlarging the atrium opening in mode B. First, the temperature on the lower floor must not be allowed to rise above a comfort limit and, second, the ventilation through this zone must not be suppressed below a healthy rate. All these can be accomplished by increasing the size of the lower floor window appropriately in response to the increase in the size of the atrium opening. Finally, the temperature on the upper floor must always be kept above that on the lower floor to prevent mode C from kicking in and bringing foul air from the upper zone to the lower zone. This can be achieved by not increasing the atrium opening excessively. However, note that, in mode C, even though there is mixing of foul air from the upper zone, the flow through the building is the fastest and the interior temperature is the lowest of the three modes. This is because, in order for this mode to develop, the atrium opening and the upper floor window need to be sufficiently large for the heat flux on the upper floor to allow the air in the upper zone to cool below the air in the lower zone and sink to the base of the space. Therefore, mode C could be desirable when large ventilative cooling was required, for example, in high summer, provided that the levels of contaminants convected from the upper floor were not excessive.

At this point it will be useful to extend the problem a little further. It is possible in certain circumstances that the lower floor contains no discernible heat load but still requires ventilation. The situation may arise when the lower floor is used as an archive, for instance, which, though containing no occupants, still needs to be kept free from excessive humidity build-up. In this case, the lower floor may be ventilated using 'borrowed' buoyancy from the upper floor – that is buoyancy generated in the upper part of the building may be used to pull fresh air through the lower floor (Fig. 5.20). In this flow regime, to enhance the ventilation through the lower floor, the height of the atrium opening or the size of the lower floor window may be increased. However, care must be taken in doing the latter, as it has side effects. Increasing the size of the lower floor window, thereby allowing more ambient air to enter the system, reduces buoyancy in the atrium. Consequently, the flow through the upper floor is suppressed and the temperature therein rises. This could cause the upper floor to become stuffy or overheated, especially in summer or when it is densely occupied. To prevent this problem, the size of the upper floor window needs to be increased appropriately relative to the increase in the size of the lower floor window.

Figure 5.21. An example of a space connected sideways. In this case, a canteen opens onto a lobby through a doorway, with each space having its own windows.

5.2.2. Spaces connected sideways

Spaces can be connected not only by being stacked on top of one another; they can also be joined sideways. A canteen may open onto a lobby through a doorway, or a meeting hall may be coupled to a foyer via a series of windows or ventilation grilles (e.g. Fig. 5.21). Often, each of these spaces also has its own heat loads and ventilation apertures which open onto the exterior. Air, therefore, can enter the building from one space and leave through the other, allowing buoyancy generated in either space to influence flow through the entire building. The resultant flow regimes are often complex, with their complexity depending chiefly on the number of spaces, openings

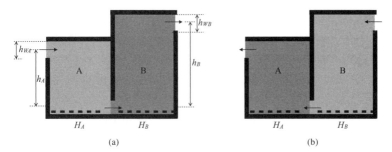

Figure 5.22. A simple system of two spaces that are connected by a low-level doorway. Each space is heated independently by a uniform source at the base and has its own window. (a) and (b) show the two steady-state ventilation regimes that can develop within the system. (Following Chenvidyakarn & Woods 2010.)

and buoyancy sources involved. However, an insight into the shared basic mechanism of these flows may be gained by considering a relatively simple system. This is shown in Fig. 5.22. In this system, two spaces, called A and B say, are connected through a doorway. Each space has its own window, which is located higher than the doorway, and which opens onto the exterior environment of uniform temperature. Each space is also heated independently by a uniform source at the base, analogous to an underfloor heater or distributed occupants.

Chenvidyakarn and Woods (2010) showed that two stable flow directions are possible in this building: one from space A to space B (denoted by A → B for convenience; Fig. 5.22a) and the other from space B to space A (denoted by B → A; Fig. 5.22b). In either of these flow regimes, the ventilation volume fluxes through both spaces are identical, thanks to mass conservation. However, the temperatures in the two spaces are different, depending on the relative position of the spaces on the streamline. The upstream space is heated by the heat load in the upstream space alone and attains a temperature above that of the exterior air. The downstream space, however, is heated by a combination of the heat load in the downstream space and heat advected from the upstream space by ventilation. Therefore, it attains a temperature above that of the upstream space. This difference in temperature, and hence buoyancy, in the two spaces drives the overall ventilation flow.

The above description may be conveniently expressed in mathematical terms. Let us denote the effective opening area of the entire system by A^* (defined according to Eq. (2.4-ii)), the height of the window in the upstream space by h_U, the height of the window in the downstream space by h_D, reduced gravity in the upstream space by $g'_U = g\alpha\Delta T_U$, and reduced gravity in the downstream space by $g'_D = g\alpha\Delta T_D$, with ΔT_U and ΔT_D being the temperature excess in the upstream and downstream space above the exterior air, respectively. An expression for the net volume flow rate Q may be obtained from tracing pressure along the streamline and applying Bernoulli's theorem. These give

$$Q = A^* \left(g'_D h_D - g'_U h_U\right)^{1/2} . \qquad (5.2.2\text{-i})$$

Expressions for the temperature in the upstream and downstream space may be obtained from the conservation of thermal energy. If the heat load in the upstream space is denoted by H_U and that in the downstream space by H_D, we have for the upstream space

$$\Delta T_U = \frac{H_U}{\rho C_P Q}, \qquad (5.2.2\text{-ii})$$

and for the downstream space

$$\Delta T_D = \frac{H_U + H_D}{\rho C_P Q}, \qquad (5.2.2\text{-iii})$$

where ρ and C_P are the density and heat capacity of air, respectively. Note that since the above equations define the spaces according to their positions on the streamline, they apply to flows in both directions.

We may readily appreciate how the possibility of multiple flow directions can affect indoor air quality and the level of comfort in the building. Suppose the two spaces represent, for instance, a kitchen connected to a living room in a house, a machine room connected to a workshop in a factory, or an office connected to a clean zone in a hospital. In any of these cases, contaminants generated in one space, whether they are in the form of cooking smoke, a leaked gas, engine fumes or germs, could spread into the other space unintentionally if the direction of flow between the two spaces were not carefully controlled. This would lead to poor indoor air quality, causing discomfort or even harming the occupants inside. Moreover, because each space attains a different temperature depending on its position on the streamline, it could be thermally comfortable or uncomfortably hot or cold, depending on the flow direction.

Therefore, to achieve thermal comfort and good indoor air quality, control of the ventilation regime is crucial. Chenvidyakarn and Woods (2010) showed that for a system of spaces with equal window heights, the flow direction depends primarily on the relative strength of the heat loads

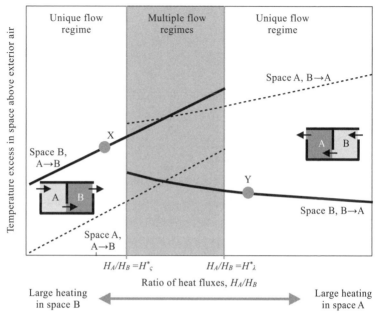

Figure 5.23. Variation in the flow regime and interior temperatures as the ratio of the heat fluxes in the two spaces changes. The dashed lines represent the temperature in space A, and the solid lines represent the temperature in space B. (Based on calculations by Chenvidyakarn & Woods 2010.)

in the spaces. This statement is captured in Fig. 5.23. In this figure, the dashed lines represent the temperature in space A, and the solid lines represent the temperature in space B. It can be seen that when the heat load in either space, H_A or H_B, is sufficiently larger than that in the other (see the zone at either end of the x-axis), flow is directed only from the weakly heated space to the strongly heated space. However, when the two heat loads are comparable (see the shaded zone in the middle), flow in either direction is possible, leading to the possibility of multiple flow regimes within a single set of boundary conditions, not dissimilar to what we have seen in Section 5.1.1 in a space equipped with multiple stacks. When multiple flow regimes are possible, the actual regime that develops depends on the history of flow. If the flow is originally from space A to space B, it is maintained in this direction until the heat load in space A is increased to a certain value in excess of that in space B (i.e., to the value $H_A/H_B = H^*_\lambda$), at which point the flow becomes unstable and changes its direction. Conversely, the flow from space B to space A is maintained until the heat load in space A is reduced to a critical value below that in space B (i.e., to the value $H_A/H_B = H^*_\zeta$). At this point, the flow from space B to space A becomes unstable and the flow from space A to space B is re-established.

What causes the flow instability and thus the change in the flow regime is as follows. At each point of regime change, the heat load in the upstream space is sufficiently large compared with that in the downstream space that the speed of the building-scale flow, which is driven by the difference in buoyancy between the two spaces, becomes so small that it is matched by the speed of exchange flow through the window on the downstream face of the building, which is driven by reduced gravity in the downstream space. Consequently, the building-scale flow is no longer able to suppress the local exchange flow, allowing the overall flow to begin to change direction and the system to evolve to a new state of equilibrium. Since the speed of the building-scale flow scales as $(g'_D h_D - g'_U h_U)^{1/2}$, and the speed of the exchange flow scales as $A_D/A^*(g'_D h_{WD})^{1/2}$, it follows that the flow regime changes when

$$(g'_D h_D - g'_U h_U)^{1/2} = \frac{A_D}{A^*}(g'_D h_{WD})^{1/2}, \quad (5.2.2\text{-iv})$$

where A_D is the area of the downstream window taking into account pressure loss across it, and h_{WD} is the vertical extent of the downstream window. Combining Eqs. (5.2.2-ii)–(5.2.2.-iv), we obtain that, for the simple case in which the windows in the two spaces are of the same height, $h_U = h_D = h_A = h_B = h$, the flow from space A to space B reverses when the heat load in space A is increased to the value such that

$$\frac{H_A}{H_B} = \left(\frac{1}{3\frac{h_{WB}}{h}}\right) - 1 = H^*_\lambda, \quad (5.2.2\text{-v})$$

where h_{WB} is the vertical dimension of the window in space B. In addition, the flow from space B to space A reverses when the heat load in space A is reduced to the value such that

$$\frac{H_A}{H_B} = \left(\frac{1}{1 - 3\frac{h_{WA}}{h}}\right) - 1 = H^*_\zeta, \quad (5.2.2\text{-vi})$$

where h_{WA} is the vertical dimension of the window in space A. Equations (5.2.2-v) and (5.2.2-vi) indicate that when the windows in the two spaces are of the same height, the critical values of the relative heat load, H^*_λ and H^*_ζ, at which the flow regime changes, depend on the size of the downstream window. This is shown graphically in Fig. 5.24. It can be seen that taller downstream windows, allowing faster exchange flow, cause the building-scale flow to change direction more readily. This reduces the value of H^*_λ but increases the value of

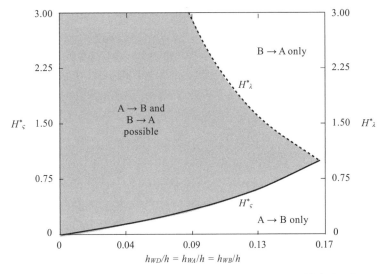

Figure 5.24. Variation in the critical heat load ratios at which the flow regime changes, H^*_ς and H^*_λ, for the system with equal window heights, as the vertical dimension of the downstream window changes. (Based on model by Chenvidyakarn & Woods 2010.)

H^*_ς, narrowing the range of heat load ratios over which multiple flow regimes may develop within a single set of boundary conditions. As the vertical dimension of the downstream window increases to about 17% of the total height of the space, $h_{WD}/h = h_{WA}/h = h_{WB}/h \approx 0.17$, the values of H^*_λ and H^*_ς coincide, so that the possibility of multiple flow regimes is eliminated, and each combination of heat loads and space geometry admits a unique flow solution. This suggests that for a building of typical height 3 m and equal window sizes, multiple flow regimes occur only when the windows are shorter than about 50 cm.

As we have already discussed, a change in the flow regime leads to a change in the relative temperature of the two spaces. A further point of interest may be noted. Compare point X and point Y in Fig. 5.23, which show variation in the temperature in downstream space B as the flow regime changes from A → B to B → A. It can be seen that as the relative heat load in the original upstream space (space A) is increased so that the flow direction changes, the temperature in the original downstream space (space B) not only becomes lower than that in the upstream space, but also decreases below its own original value. This indicates that, in a system of laterally connected spaces, it is possible for *an increase in the heat load in one space to cause the other space to cool down.* This may seem counter-intuitive, as one would normally expect an increase in any heat load to lead to an increase in temperature, not a decrease; but it is a consequence of a change in the heat balance of the system as the flow direction changes, as described earlier.

The above phenomenon is not just an academic curiosity; it has important practical implications. Consider this example. Suppose that space A has a smaller heat load than space B, so that at steady state flow is directed from space A to space B. On a cold day, it might be possible that a combination of heat inputs from the occupants and heating system in space B was sufficient to keep the temperature in that space to within a comfortable range, whereas a combination of heat inputs from similar sources in space A was insufficient to maintain thermal comfort within it. In this situation, a common reaction of the occupants in space A would be to turn up the heating in their space in order to raise the temperature within it to a comfortable value. If this led to the total heat load in space A reaching a certain critical value above that in space B, the flow direction could reverse unbeknownst to the occupants in both spaces. The new flow regime, directed from space B to space A, could then cause the temperature in space B to fall below a comfortable value, inducing the occupants in space B to also turn up the heating within it. A ratchet mechanism would kick in: any increase in the heating in space B beyond a certain critical value above that in space A would reverse the flow direction again, causing the temperature in space A to fall and the occupants in space A to increase the heating in their space even further. This 'heating race' would likely continue until discomfort was felt in either space due to excessive heating. The result would be a waste of energy.

A more appropriate operational scheme for the above situation would be one based on an understanding of flow mechanics outlined earlier. When space A is uncomfortably cold, the size of the window in space B could be reduced (by means of a building automation system, for instance), so as to reduce the net flow through the system and raise the temperature in both spaces. This would allow the heating in space A to be increased by a lesser amount than would otherwise be required, and the heating in space B to perhaps be reduced from its original value. This would not only save energy, but also make the flow regime easier to control, since the reduction in the size of the window on the downstream face of the building would make it more difficult for the flow to change direction.

The aforementioned flow principles carry through to the case of connected spaces with unequal window heights, as in, for example, a single-storey office connected to a tall atrium. In this case, Eqs. (5.2.2-i)–(5.2.2-iv) still apply. However, the condition for regime change now becomes dependent on the relative height of the windows, as well as the relative heat load and the relative size of the downstream windows. Formal expressions for the limits of each flow regime may be obtained, as in the case of the system with equal vent heights, from applying to the heat balance equations (Eqs. (5.2.2-ii) and (5.2.2-iii)) the condition for an exchange flow to develop at the downstream window (Eq. (5.2.2-iv)). This will show that, if space A is shorter than space B, the flow from space A to space B exists only when the window in space B is sufficiently high or when the heat load in space A is sufficiently small. In a simple case in which the sizes of the windows and the door are equal, this condition is met when

$$\frac{h_A}{h_B} < \frac{\left(\frac{H_A}{H_B} + 1\right)\left(1 - 3\frac{h_{WB}}{h_B}\right)}{\left(\frac{H_A}{H_B}\right)}. \qquad \text{(5.2.2-vii)}$$

Furthermore, the flow is directed from space B to space A when the heights of the two windows are comparable and the heat load in space A (which is the shorter of the two spaces) is sufficiently large. If the sizes of the two windows and the door are equal, this condition is satisfied when

$$\frac{h_A}{h_B} > \frac{3h_{WA}}{h_B} + \frac{1}{\left(\frac{H_A}{H_B} + 1\right)}. \qquad \text{(5.2.2-viii)}$$

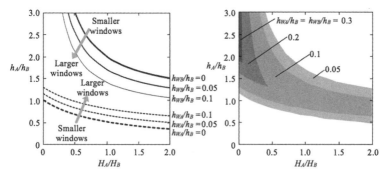

Figure 5.25. Limits of the flow regimes that develop in a system of two connected spaces with unequal window heights. For each window size, the flow from space A to space B exists below the corresponding solid line, and the flow from space B to space A exists above the corresponding dashed line (a). The range of combinations of window heights and heat loads that allows multiple flow regimes to develop within a single set of boundary conditions is bounded by each pair of solid and dashed lines in (a) and described by each shaded zone in (b). Both plots are for a simple case where the areas of the windows and the connecting doorway are equal. (Adapted from Chenvidyakarn & Woods 2010.)

Thus, multiple flow regimes are possible within the same set of boundary conditions only when the relative height of the windows lies within the range

$$\frac{3h_{WA}}{h_B} + \frac{1}{\left(\frac{H_A}{H_B}+1\right)} < \frac{h_A}{h_B} < \frac{\left(\frac{H_A}{H_B}+1\right)\left(1-3\frac{h_{WB}}{h_B}\right)}{\left(\frac{H_A}{H_B}\right)} \qquad (5.2.2\text{-ix})$$

These limits of the flow regimes are shown in Fig. 5.25. The ranges of geometrical values and heat loads which allow multiple flow regimes are represented by the shaded zones in Fig. 5.25b. It can be seen that multiple flow regimes become more difficult to establish (i.e., the shaded zone gets smaller) as the vertical dimension of the window in the downstream space increases. This is because larger windows allow stronger exchange flows, increasing the tendency for the overall flow to change direction.

To wrap up this section, it will be useful to discuss a little what happens if only one of the spaces is heated. In this situation, the flow regimes that can develop are quite different from those discussed in the preceding text, and multiple flow directions may or may not be possible, depending primarily on the positions of the openings and the heat source. For a short space connected to a tall one through a low-level opening (e.g., a single-storey office linked to a multi-storey atrium by a doorway), if the taller space alone is heated, only one flow direction is possible; that is from the shorter to taller space (Fig. 5.26). The shorter space acts effectively as an inlet chamber for the taller space so that the system behaves as if it consisted of the taller space alone. The interior temperature structure in the taller space depends on the type of heat load present. If the heat load is uniformly distributed, it becomes well-mixed at a temperature above that of the exterior air at steady state (Fig. 5.26a). However, if the heat load is localised, it becomes stratified, with the upper zone attaining a temperature above that of the exterior air and the lower zone maintained at the ambient air temperature (Fig. 5.26b). In both cases, the shorter space is filled with air at the ambient temperature.

In contrast, if the shorter space alone is heated, several flow directions are possible (cf. Fitzgerald & Woods 2008). If the heat load is sufficiently small, flow is directed from the shorter to taller space (Fig. 5.27a). If the heat load is

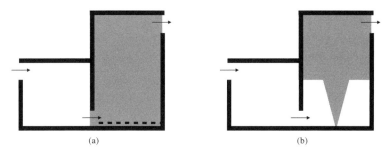

Figure 5.26. Steady-state flow regimes that develop in a system of spaces of unequal height connected at a low level, when the taller space alone is heated by a distributed source (a), and when it alone is heated by a localised source (b). (Based on Gladstone & Woods 2001 and Linden, Lane-Serff & Smeed 1990.)

sufficiently large, flow is directed in the opposite direction (Fig. 5.27b). The interior temperature structure depends on the flow regime that develops. When the flow is from the shorter to taller space, if the source is distributed, both spaces become well-mixed and attain a uniform temperature above

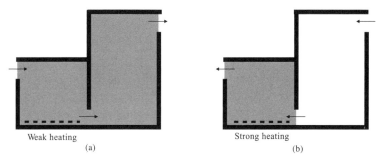

Weak heating   Strong heating
        (a)   (b)

Figure 5.27. Steady-state flow regimes which develop in a system of spaces of unequal height connected at a low level, when the shorter space alone is heated by a distributed source. (Based on Fitzgerald & Woods 2008.)

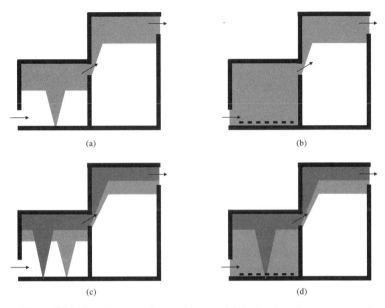

Figure 5.28. Steady-state flow regimes which develop in a system of spaces of unequal height connected at a high level, when the shorter space alone is heated by a localised source (a), a distributed source (b), multiple localised sources of unequal strength (c) and a combination of localised and distributed sources. (Based on Holford & Hunt 2003, Linden & Cooper 1996, Cooper & Linden 1996, and Fitzgerald & Woods 2008.)

that of the exterior air, whereas if the flow is from the taller to shorter space, only the shorter space is heated and becomes well-mixed. The taller space acts as an inlet chamber for the shorter space and attains the temperature of the exterior air.

The flow picture will be quite different if the two spaces are connected instead by a high-level opening, for example, a ventilation grille. With the shorter space being the only one heated, the possibility of multiple flow directions is eliminated; the taller space becomes effectively a ventilation stack for the shorter space. If the shorter space contains a localised heat source, both spaces become stratified (Fig. 5.28a; Holford & Hunt 2003). However, if the shorter space contains a distributed source, the shorter space becomes well-mixed but the taller space is stratified (Fig. 5.28b). In the case in which the shorter space contains multiple localised sources of unequal strength or a combination of localised and distributed sources, complex vertical stratification develop in both spaces along the lines of Figs. 5.28c and d (Cooper & Linden 1996; Linden & Cooper 1996; Fitzgerald & Woods 2008).

Final remarks

It has been demonstrated over the course of this book how variation in the geometry of the space and the strength of cooling/heating can lead to different flow regimes, flow rates and interior temperatures, and hence different levels of comfort and energy efficiency. From this, we can appreciate how essential the control of ventilation features and buoyancy forces are to effective natural ventilation.

The operation of naturally ventilated buildings will likely become more sophisticated as our expectations of building performance increase. We only have to look at the differences between traditional buildings and their modern counterparts to be convinced of this. Whereas the former need only to utilise simple manual controls to keep indoor temperatures and ventilation rates within relatively relaxed ranges, the latter increasingly have to rely on complicated computerised management systems to regulate internal conditions to meet stringent standards of comfort and energy efficiency. Undoubtedly, our knowledge of how to control naturally ventilated buildings will advance as more information from the field becomes available on specific flow phenomena and occupant behaviour. We can be confident, however, that whatever developments there will be in the future, the fundamental effects of buoyancy discussed herein will always be relevant. It is only by having a firm grasp of basic flow processes and then building on this knowledge that we can hope to achieve successful, comfortable and efficient naturally ventilated buildings.

References

Alamdari, F, Butler, DJG, Grigg, PF & Shaw, MR. 1998. Chilled ceilings and displacement ventilation. *Renewable Energy*, 15: 300–5.

Allard, F (Editor). 1998. *Natural ventilation in buildings: A design handbook*. London: James & James.

Arnold, D. 2004. Underfloor air conditioning in North America. Presentation slide. CIBSE/ASHRAE Group Meeting, London South Bank University, London, 8 December 2004.

Awbi, HB. 1991. *Ventilation of buildings*. London and New York: Chapman & Hall.

Baines, WD. 1975. Entrainment by a plume or jet at a density interface. *Journal of Fluid Mechanics*, 68: 309–20.

Baines, WD & Turner, JS. 1969. Turbulent buoyant convection from a source in a confined region. *Journal of Fluid Mechanics*, 37: 51–80.

Baines, WD, Turner, JS & Campbell, IH. 1990. Turbulent fountains in an open chamber. *Journal of Fluid Mechanics*, 212: 557–92.

Batchelor, GK. 1954. Heat convection and buoyancy effects in fluids. *Quarterly Journal of the Royal Meteorological Society*, 80: 339–58.

Bloomfield, LJ & Kerr, RC. 2000. A theoretical model of a turbulent fountain. *Journal of Fluid Mechanics*, 424: 197–216.

Bower, DJ, Caulfield, CP, Fitzgerald, SD & Woods, AW. 2008. Transient ventilation dynamics following a change in strength of a point source of heat. *Journal of Fluid Mechanics*, 614: 15–37.

Brown, WG & Solvason, KR. 1962a. Natural convection heat transfer through rectangular openings in partitions I. *International Journal of Heat and Mass Transfer*, 5: 859–68.

Brown, WG & Solvason, KR. 1962b. Natural convection heat transfer through rectangular openings in partitions II. *International Journal of Heat and Mass Transfer*, 5: 869–78.

Brown, WG, Wilson, AG & Solvason, KR. 1963. Heat and moisture flow through openings by convection. *Journal of the American Society of Heating, Refrigerating and Air-Conditioning Engineers*, 5: 49–54.

Castaing, B, Gunaratne, G, Heslot, F, Kadanoff, L, Libchaber, A, Thomae, S, Wu, X-Z, Zaleski, S & Zanetti, G. 1989. Scaling of hard turbulence in Rayleigh–Bénard convection. *Journal of Fluid Mechanics*, 204: 1–30.

Chen, Q & Glicksman, L. 2001. Application of computational fluid dynamics for indoor air quality studies. In *Air quality handbook*. New York: McGraw-Hill.

Chen, ZD & Li, Y. 2002. Buoyancy-driven displacement natural ventilation in a single-zone building with three-level openings. *Building and Environment*, 37: 295–303.

Chenvidyakarn, T & Woods, AW. 2004. The control of pre-cooled natural ventilation. *Building Services Engineering Research and Technology*, 25(2): 127–40.

Chenvidyakarn, T & Woods, AW. 2005a. Multiple steady states in stack ventilation. *Building and Environment*, 40(3): 399–410.

Chenvidyakarn, T & Woods, AW. 2005b. Top-down pre-cooled natural ventilation. *Building Services Engineering Research and Technology*, 26(3): 181–93.

Chenvidyakarn, T & Woods, AW. 2006. Stratification and oscillations produced by pre-cooling during transient natural ventilation. *Building and Environment*, 42(1): 99–112.

Chenvidyakarn, T & Woods, AW. 2008. On underfloor air-conditioning of a room containing a distributed heat source and a localised heat Source. *Energy and Buildings*, 40: 1220–7.

Chenvidyakarn, T & Woods, AW. 2010. On the natural ventilation of two independently heated spaces connected by a low-level opening. *Building and Environment*, 45(3): 586–95.

Churchill, SW & Ozoe, H. 1973. Correlations for laminar forced convection in flow over an isothermal flat plate and in developing and fully developed flow in an isothermal tube. *Journal of Heat Transfer*, 95: 46.

Cook, MJ, Lomas, KJ & Eppel, H. 1999. Design and operating concept for an innovative naturally ventilated library. In *Proceedings of engineering in the 21st century: The changing world*. CIBSE National Conference, London, 4–5 October 1999, pp. 500–7.

Cooper, P & Linden, PF. 1996. Natural ventilation of an enclosure containing two buoyancy sources. *Journal of Fluid Mechanics*, 311: 153–76.

Cooper, P & Mak, N. 1991. Thermal stratification and ventilation in atria. In *Proceedings of Australian and New Zealand Solar Energy Society Conference*, Adelaide, Australia, pp. 385–91.

Crisp, VHC, Fisk, DJ & Salvidge, AC. 1984. *The BRE Low-Energy Office*. Department of the Environment, Building Research Establishment, Watford, UK.

Dalziel, SB &Lane-Serff, GF. 1991. The hydraulics of doorway exchange flows. *Building and Environment*, 26(2): 121–35.

Davies, GMJ. 1993. *Buoyancy driven flow through openings*. PhD thesis, Cambridge University, UK.

Deardorff, JW, Willis, GE & Lilly, DK. 1969. Laboratory investigation of non-steady penetrative convection. *Journal of Fluid Mechanics*, 35: 7–31.

Denton, RA & Wood, IR. 1979. Turbulent convection between two horizontal plates. *International Journal of Heat and Mass Transfer*, 22: 1339–46.

Epstein, M. 1988. Buoyancy-driven exchange flow through openings in horizontal partitions. *The international conference on vapor cloud modelling*, Cambridge, MA, 2–4 November 1987.

Etheridge, DW & Sandberg, M. 1996. *Building ventilation: Theory and measurement*. New York: John Wiley & Sons.

Fisk, DJ. 1981. *Thermal control of buildings*. London: Applied Science.

Fitzgerald, SD & Woods, AW. 2004. Natural ventilation of a room with vents at multiple levels. *Building and Environment*, 39: 505–521.

Fitzgerald, SD & Woods, AW. 2007. Transient natural ventilation of a room with a distributed heat source. *Journal of Fluid Mechanics*, 591: 21–42.

Fitzgerald, SD & Woods, AW. 2008. The influence of stacks on flow patterns and stratification associated with natural ventilation. *Building and Environment*, 43: 1719–33.

Fujii, T & Imura, H. 1972. Natural convection heat transfer from a plate with arbitrary inclination. *International Journal of Heat and Mass Transfer*, 15: 755.

Garon, AM & Goldstein, RJ. 1973. Velocity and heat transfer measurements in thermal convection. *Physics of Fluids*, 16: 1818–25.

Germeles, AE. 1975. Forced plumes and mixing of liquids in tanks. *Journal of Fluid Mechanics*, 71: 601–23.

Gladstone, C & Woods, AW. 2001. On buoyancy-driven natural ventilation of a room with a heated floor. *Journal of Fluid Mechanics*, 441: 293–314.

Gorton, RL & Sassi, MM. 1982. Determination of temperature profiles in a thermally stratified air-conditioned system: Part 2. Program description and comparison of computed and measured results. *Transactions of the American Society of Heating, Refrigerating and Air-conditioning Engineers.* 88(2): paper 2701.

Holford, JM & Hunt, G. 2003. Fundamental atrium design for natural ventilation. *Building and Environment*, 38: 409–26.

Holford, JM & Woods, AW. 2007. On the thermal buffering of naturally ventilated buildings through internal thermal mass. *Journal of Fluid Mechanics*, 580: 3–29.

Holman, JP. 1997. *Heat transfer* (8th edition). New York: McGraw-Hill.

Hunt, GR, Holford, JM & Linden, PF. 2001. Natural ventilation by competing effects of localised and distributed heat sources. In *Proceedings of the 14th Australasian Fluid Mechanics Conference*, Adelaide, Australia, 9–14 December 2001, pp. 545–85.

Hunt, GR & Linden, PF. 1997. Displacement and mixing ventilation driven by opposing wind and buoyancy. *Journal of Fluid Mechanics*, 527: 27–55.

Jacobsen, J. 1988. Thermal climate and air exchange rate in a glass covered atrium without mechanical ventilation related to simulations. In *Proceedings of the 13th national solar conference*, Cambridge, MA, 18 June 1988, 4: 61–71.

Kaye, N. 1998. *Interaction of turbulent plumes*. PhD thesis, Cambridge University, UK.

Kaye, NB & Hunt, GR. 2004. Time-dependent flows in an emptying filling box. *Journal of Fluid Mechanics*, 520: 135–56.

Kenton, AG, Fitzgerald, SD & Woods, AW. 2004. Theory and practice of natural ventilation in a theatre. In *Proceedings of the 21th passive and low energy architecture conference*, Eind-hoven, The Netherlands, 19–22 September 2004.

Keil, DE. 1991. *Buoyancy driven counterflow and interfacial mix-ing*. PhD thesis, Cambridge University, UK.

Kumagai, M. 1984. Turbulent buoyant convection from a source in a confined two-layered region. *Journal of Fluid Mechanics*, 147: 105–31.

Lane-Serff, GF. 1989. *Heat flow and air movement in buildings*. PhD thesis, Cambridge University, UK.

Lane-Serff, GF, Linden, PF & Simpson, JE. 1987. Transient flow through doorways produced by temperature differences. In *Proceedings of ROOMVENT '87*, Stockholm, Sweden, 10–12 June 1987, pp. 41–52.

Lee, JH-W & Chu,V. 2003. *Turbulent jets and plumes: A Lag-rangian approach*. New York: Springer.

Li, Y & Yam, JCW. 2004. Designing thermal mass in naturally ventilated buildings. *International Journal of Ventilation*, 2: 313–24.

Lin, YJP & Linden, PF. 2005. The entrainment due to a turbulent fountain at a density interface. *Journal of Fluid Mechanics*, 542: 25–52.

Linden, PF. 1973. The interaction of a vortex ring with a sharp density interface: A model for turbulent entrainment. *Journal of Fluid Mechanics*, 60: 467–80.

Linden, PF. 1999. The fluid mechanics of natural ventilation. *Annual Review of Fluid Mechanics*, 31: 201–38.

Linden, PF, Lane-Serff, GF & Smeed, DA. 1990. Emptying filling boxes: The fluid mechanics of natural ventilation. *Journal of Fluid Mechanics*, 212: 309–35.

Linden, PF & Simpson, JE. 1985. Buoyancy driven flows through an open door. *Air Infiltration Review*, 6: 4–5.

Lishman, B & Woods, AW. 2006. The control of naturally ventil-ated buildings subject to wind and buoyancy. *Journal of Fluid Mechanics*, 557: 451–71.

Lister, JR. 1995. On penetrative convection at low Péclet number. *Journal of Fluid Mechanics*, 292: 229–48.

Livermore, SR & Woods, AW. 2006. Natural ventilation of multiple storey buildings: The use of stacks for secondary ventilation. *Building and Environment*, 41: 1339–51.

Livermore, SR & Woods, AW. 2007. Natural ventilation of a building with heating at multiple levels. *Building and Environment*, 42: 1417–30.

Livermore, SR & Woods, AW. 2008. On the effect of distributed cooling in natural ventilation. *Journal of Fluid Mechanics*, 600: 1–17.

Long, K. 2001. Underneath the arches. *Building Design*, 12 April 2001: 15–17.

McDougall, TJ. 1981. Negatively buoyant vertical jets, *Tellus*. 33: 313–20.

Mingotti, N, Chenvidyakarn, T & Woods, AW. 2011. The fluid mechanics of the natural ventilation of a narrow-cavity double-skin facade. *Building and Environment*, 46: 807–23.

Morton, BR, Taylor, G & Turner, JS. 1956. Turbulent gravitational convection from maintained and instantaneous sources. *Proceedings of Royal Society of London A*, 234: 1–23.

Neufert, E. 1980. *Architects' data*, Second (international) English edition (General Editor: Jones, V). Oxford: Blackwell Science.

Novoselac, A & Srebric, J. 2002. A critical review on the performance and design of combined cooled ceiling and displacement ventilation systems. *Energy and Buildings*, 34: 497–509.

Rooney, GG & Linden, PF. 1996. Similarity considerations for non-Boussinesq plumes in an unstratified environment. *Journal of Fluid Mechanics*, 318: 237–50.

Rouse, H, Yih, C-S & Humphreys, HW. 1952. Gravitational convection from a boundary source. *Tellus*, 4: 201–10.

Savardekar, K. 1990. *Aspects of passive cooling: A study on natural ventilation*. MPhil thesis, Cambridge University, UK.

Schmidt, W. 1941. Turbulent propagation of a stream of heated air. *Zeitschrift für Angewandte Mathematik und Mechanik*, 21: 265 and 351.

Shaw, BH & Whyte, W. 1974. Air movement through doorways: The influence of temperature and its control by forced air flow. *Journal of the Institute of Heating and Ventilating Engineers*, 42: 210–18.

Thomas, PH, Hinkley, PL, Theobald, CR & Simms, DL.1963. Investigations into the flow of hot gases in roof venting. *Fire Research Technical Paper 7*. Fire Research Station, Watford, UK.

Townsend, AA. 1959. Temperature fluctuations over a heated horizontal surface. *Journal of Fluid Mechanics*, 5: 209–41.

Townsend, AA. 1970. Entrainment and the structure of turbulent flow. *Journal of Fluid Mechanics*, 41: 13–46.

Turner, JS. 1973. *Buoyancy effects in fluids*. Cambridge, UK: Cambridge University Press.

Woods, AW, Caulfield, CP & Phillips, JC. 2003. Blocked natural ventilation: The effect of a source mass flux. *Journal of Fluid Mechanics*, 495: 119–33.

Worster MG & Huppert HE. 1983. Time-dependent density profiles in a filling box. *Journal of Fluid Mechanics*, 132: 457–66.

Zilintikevich, SS. 1991. *Turbulent penetrative convection*. Aldershot, UK: Avebury.

Index

air-borne contaminants, spread
    and control of, 210, 226, 233
air-conditioning
  as compared with passive
    cooling, 128–31
  as generator of cold mass
    plume, 58, 76–7
  control of, 69, 75–6, 187–8
  energy consumption of, 130
  reduced popularity of, 1
  underfloor, 116

Bernoulli's theorem, description
    of, 26–7
Boussinesq approximation,
    definition of, 10–11
Boussinesq plume, basic
    behaviour of, 55–8
buoyancy
  definition of, 2
  modelling of, 4, 5–8
buoyancy flux
  conservation of, 21–3
  definition of, 21, 22

  of flow in stratified
    environment, 22–3, 57, 71,
    87–8, 186
  scaling of, 14

characteristic length scale,
    in buoyancy-driven flow,
    12
chilled ceiling, 97, 165, 170, 172–4,
    180–2
cold feet/leg sensation, 103, 153,
    188
computational fluid dynamics
    (CFD), 8, 16
conductive heat transfer,
    treatment of, in water-bath
    modelling, 14–16
connected spaces
  laterally, 230–45
  vertically (*see* multi-storey
    buildings)
cooling
  external/pre-, 127–65
  internal, 165–89

convective heat transfer,
   treatment of, in water-bath
   modelling, 14–15

density of fluid, measurement of,
   7
diffusivity, definition of, 11
displacement ventilation,
   definition of, 45
distributed source of buoyancy
   interaction of, with localised
      source of buoyancy,
      116–25
   water-bath modelling of,
      7–8
double-skin façade, 193–4
draining ventilation (*see* flushing
   ventilation)
dynamic similarity, 5, 9, 12–14

effective opening area, definition
   of, 27–9
entrainment
   in plume, 52, 54
   penetrative (*see* penetrative
      convection)
equilibrium, hydrostatic,
   definition of, 23
exchange flow
   across non-vertical openings,
      42–3, 143
   across vertical openings, 38,
      40–2, 187, 235–6, 240–2
   definition of, 38
   elimination of, using stack,
      142–3

filling-box process, 61

fire, ventilation of, 61, 82–3
flushing ventilation, 35–50,
   127–43
Froude number, definition of, 78,
   91–2
Full-scale flow modelling, 4, 16

gas leak, ventilation of, 49–50,
   82–3
global warming, 1

heated floor (*see* distributed
   source of buoyancy)
hydrostatic equilibrium (*see*
   equilibrium)
hydrostatic pressure
   basic profiles of, in cold and
      warm rooms, 24–5
   description of, 23

instability
   caused by pre-cooling, 131–5,
      158–61
   in flow through connected
      spaces, 234–7, 240–2
   Rayleigh-Taylor, 23, 42, 133
intermediate/middle vent, impact
   of, 204–18

kinematic viscosity, definition of,
   11

localised source of buoyancy
   interaction of, with distributed
      source of buoyancy, 116–25,
      175–83
   water-bath modelling of, 6–7,
      58–9

mass, conservation of, 18–19
mechanical ventilation, reduced
    popularity of, 1
mixing ventilation, definition of,
    38
multiple flow regimes
    in laterally connected spaces,
        231–44
    in space with multiple stacks,
        191–204
multi-storey buildings, ventilation
    of, 218–29

natural ventilation
    energy-saving potential of,
        130–1
    history of, 1
neutral level
    definition of, 24
    effects of, on flow direction,
        24–5
night cooling/purging, 174–5,
    182–3
non-Boussinesq flow, 11
non-monotonic behaviour, in
    flow driven by pre-cooling,
    149–53

oscillation, in flow driven by
    pre-cooling, 131–6

Péclet number, 8, 9, 12–13
penetrative
    convection/entrainment, 73,
    77–8, 89–92, 107–11, 113–15,
    120–2, 176–8, 186
plume
    in uniform environment, 55–7

in stratified environment, 57
theory, 51–8
virtual source of (see virtual
    plume source)
pre-cooling (see cooling,
    external)
pre-heating, 154–5
pressure
    hydrostatic (see hydrostatic
        pressure)
pressure loss coefficient
    definition of, 27–9
    values of, 41, 43, 46

radiative heat transfer, in
    water-bath modelling, 14–16
Rayleigh number, 9, 12–14
Rayleigh-Taylor instability (see
    instability, Rayleigh-Taylor)
reduced gravity, definition of,
    9–10
residual buoyancy/heat
    flows driven by (see flushing
        ventilation)
    water-bath modelling of,
        36
Reynolds number, 8, 9, 12–13

salt-bath modelling
    applicability of, 14–16
    description of, 5–8
    history of, 4–5
    theoretical basis of, 9–14
shadowgraph imagery, use in
    water-bath modelling, 7
sick building syndrome, 1
specific heat capacity, definition
    of, 20

stacks
    as generator of multiple flow
       regimes, 191–204
    use of, in real-life buildings,
       2, 3, 4, 63, 129, 130, 144,
       191–2

thermal energy, conservation of,
    20–1
thermal mass
    behaviour of, 166–8, 174–5,
       182–3
    use of, in real-life buildings,
       128–31, 165–6
turn-around time of occupancy,
    management of, 35, 40–1, 43,
       47–50, 135–6, 138–9, 141–2

underfloor heating (*see*
    distributed source of
    buoyancy)

velocity of flow
    measurement of, 7
    scaling of, 14
volume expansion coefficient,
    definition of, 10
virtual plume source
    definition of, 56–7
    in flow in stratified
       environment, 57

water-bath modelling (*see*
    salt-bath modelling)
wind force, modelling of, 4, 8